Fish Ecology

TERTIARY LEVEL BIOLOGY

A series covering selected areas of biology at advanced undergraduate level. While designed specifically for course options at this level within Universities and Polytechnics, the series will be of great value to specialists and research workers in other fields who require knowledge of the essentials of a subject.

Recent titles in the series

TERTIARY LEVEL BIOLOGY

Fish Ecology

R.J. WOOTTON

Department of Biological Sciences
University College of Wales
Aberystwyth

Blackie

Glasgow and London

Published in the USA by
Chapman and Hall
New York

Blackie and Son Limited
Bishopbriggs, Glasgow G64 2NZ
and
7 Leicester Place, London WC2H 7BP

Published in the USA by
Chapman and Hall
a division of Routledge, Chapman and Hall, Inc.
29 West 35th Street, New York, NY 10001-2291

British Library Cataloguing in Publication Data

Wootton, R.J.
    Fish ecology. — (Tertiary level biology)
    I. Title II. Series
    597.092

    ISBN 978-0-216-93151-0
    ISBN 978-0-216-93152-7 pbk

Library of Congress Cataloging-in Publication Data

Wootton, R.J. (Robert J.)
    Fish ecology / R.J. Wootton.
        p.   cm. — (Tertiary level biology)
    Includes bibliographical references and references.
    ISBN 978-0-412-02921-9 (cloth).—ISBN 978-0-412-02931-8 (paper)
    1. Fishes—Ecology.   I. Title. II. Series.
QL639.8.W66 1992
597'.05—dc20   91-28996
                                          CIP

Typeset by Thomson Press (India) Limited, New Delhi, India

Preface

Fishes live in a world that is unfamiliar to us. Although we may make brief visits to this other world using a snorkel, scuba or even more advanced diving equipment, we can never become a part of it. Yet, an understanding of fish ecology requires an awareness of the relationships between fishes and their environment. The purpose of this book is to introduce the ecology of fishes by describing the inter-relationships between fishes and the aquatic habitats they occupy. The book can be read in complementary ways. A sequential reading, chapter by chapter, covers the main themes of ecology, including habitat use, species interactions, migration, feeding, population dynamics and reproduction in relation to the major habitats occupied by fishes. An alternative reading selects a particular sort of habitat, such as rivers, and, by using the index and skipping from chapter to chapter, builds up a picture of the ecology of fishes living in that habitat.

The text is written for advanced students. Its emphasis is on descriptive rather than quantitative ecology. It is assumed that the reader will be familiar with the basic biology of fishes, acquired from a text such as *The Biology of Fishes* (Bone and Marshall, 1982) also published in the *Tertiary Level Biology* series.

I would like to thank Dr J.D. Fish and two anonymous reviewers who, within a tight time-schedule, tried to improve the text. Any mistakes and shortcomings are my contribution. I would also like to thank Denise Nicholls for the more complex line drawings. Other illustrations were prepared using 'VP-Graphics'.

R.J.W.

Contents

4 MIGRATION, TERRITORIALITY AND SHOALING IN FISHES   77

5 FEEDING AND GROWTH   98

CHAPTER ONE

THE ENVIRONMENT, ORGANISMS AND RELATIONSHIPS

1.1 Introduction

A famous poster shows earthrise seen from the moon. The beautiful image of the blue oceans and white clouds emphasizes that, despite its name, Earth is a planet of water. Most of the evolutionary history of life has been played out in the medium of water. The interactions between organisms and their abiotic environment and between organisms themselves have provided the dynamic for evolution by natural selection. Ecology is the study of those interactions. Its central question is what abiotic and biotic conditions are required for organisms to successfully reproduce themselves? The problems in ecology come not from the difficulties of the questions but from the complexities of the potential answers.

Three groups of vertebrates have played out their evolutionary history in water (Nelson, 1984). In everyday language they are called fishes, although this hides their long, separate evolutionary histories. The groups are the Agnatha, the Chondrichthyes and the Teleostomi (but also includes the terrestrial descendants of the bony fishes). The agnathans are primitive vertebrates that lack jaws. Today they are represented only by the eel-like hagfishes and lampreys. The two other groups are Gnathostomata, that is those vertebrates with jaws. The familiar sharks and rays, commonly called the elasmobranchs, and the less well-known chimaeras form the Chondrichthyes. They are cartilaginous fishes characterized by the lack of bone in their skeletons. The Teleostomi (or Osteichthyes) are jawed fishes with bony skeletons. They include the lungfishes (Dipneusti), the coelocanth (Crossopterygii) and the ray-finned fishes (Actinopterygii). This latter group is dominated by the evolutionary advanced bony fishes, the Teleostei. An understanding of the ecology of all these fishes depends on a knowledge of the properties of the medium in which they live.

1.2   Properties of water

Our familiarity with water often leads us to forget what a curious substance it is, yet its peculiarities have important consequences for the organisms that live in it—including the fishes. Many of its properties are influenced by temperature and by the quantity of dissolved inorganic material (Colt, 1984; Smith, 1976).

1.2.1   *Density, viscosity and heat capacity*

Water is about 775 times denser than air. The density of living tissue is close to that of water so it is a buoyant medium. In comparison with fully terrestrial organisms, aquatic animals do not need to develop strong skeletons to counteract the effect of gravity. The buoyancy provided by water means that movement in all three dimensions is relatively easy for well-muscled organisms such as fishes. Even fishes that usually live on the bottom will make forays up into the water column to catch prey or take advantage of water currents (see chapter 4).

The density of water is a function of temperature, salinity and, to a slight extent, depth. The effects of temperature and salinity on density are important because they impose subtle structures on aquatic environments and the ecology of fishes is affected by these structures (chapter 2).

1.2.1.1   *Density and temperature.*   Ice is less dense than liquid water at 0°C and so floats. In cold climates, a permament or seasonal layer of ice and snow isolates the liquid water below from the atmosphere above. The layer attentuates the light and restricts the diffusion of gases between the water and the atmosphere.

The density of fresh water is a maximum at about 4°C (Figure 1.1a). Consequently, both cooler and warmer water will float on a layer of water at 4°C. Above this temperature, the density decreases with an increase in temperature. The relative change in density per degree rise in temperature increases with temperature. A consequence is that it takes more work to mix waters that differ in temperature when water temperatures are high. Fishes can experience their habitat as a series of layers of water at different temperatures, the positions of these layers being maintained because of the differences in densities.

1.2.1.2   *Density and salinity.*   Waters that differ in salinity also differ in density (Figure 1.1b). Water of low salinity floats on water of a higher

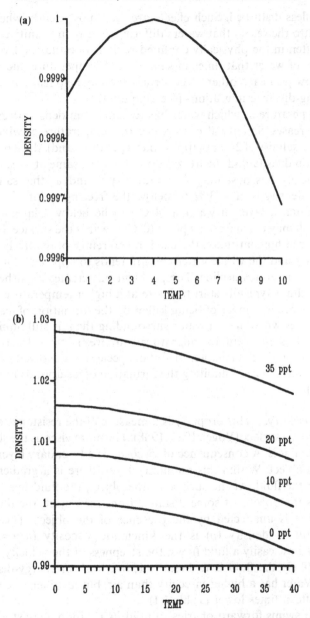

Figure 1.1 Effect of temperature and salinity on density of water: (a) effect of temperature on density of fresh water over range 0–10°C; (b) effect of temperature (0–40°C) and salinity (0–35 ppt).

salinity unless disturbed. Such effects are seen most clearly where rivers flow out into the sea so that waters differing greatly in salinity come into contact, often in the physically confined regions of estuaries. Even in the sea, bodies of water that differ in salinity and temperature and hence in density, flow past each other. Such structures may help fishes to orientate during long-distance migrations (see chapter 4).

The temperature at which water has its maximum density decreases as salinity increases. Salinity also decreases the temperature at which water freezes. At a salinity of 24.7 ppt (ppt = parts per thousand), the temperature of maximum density and the freezing point are the same at $-1.33°C$. Sea water typically has a salinity of about 35 ppt and at this salinity the maximum density is at $-3.52°C$, below the freezing point of $-1.91°C$. In fresh water, a layer of water at 4°C can lie below a layer of colder water, which may reach freezing point (0°C) towards the surface. However, in sea water at high latitudes, there can be no refuge of relatively warmer, denser water and the whole water column tends to approach the freezing point as temperatures decline. The problem this creates for fishes is that their body fluids typically start to freeze at a higher temperature than sea water. They are in danger of being killed by the formation of ice crystals in their tissues while the sea water surrounding them is still liquid. Some fish evade this problem by migrating into fresh water to avoid low temperatures in the sea (chapter 4). Other species have evolved substances that act as anti-freezes, inhibiting the formation of ice crystals in the body (chapter 2).

1.2.1.3 *Viscosity.*   This property is a measure of the resistance of a fluid to flow across a surface (Vogel, 1981, 1988). The more viscous the substance the less 'fluid' it is. A consequence of viscosity is a boundary layer of fluid around an object. Within this boundary layer, there is a gradient in the velocity of the fluid. At the surface of the object, the fluid has the same velocity as the object. At some distance from the object, the fluid has a velocity that is unaffected by the presence of the object. The ratio of viscosity (μ) to density (ρ) is the kinematic viscosity ($\mu/\rho = v$). This determines how easily a fluid flows, the steepness of the velocity gradient and how likely the fluid is to break out in a rash of energy-dissipating vortices. Water has a higher viscosity than air, but its kinematic viscosity is about fifteen times lower (Table 1.1).

As a fish swims forward or tries to hold its position against a current, it generates a thrust (Bone and Marshall, 1982; Webb, 1975). The forward motion produced by the thrust is opposed by drag, which has two

Table 1.1 Comparison of viscosity of air, fresh water and sea water. (Data from Vogel, 1988.)

	Viscosity (Pa.s)	Density (kg m^{-3})	Ratio (kinematic viscosity) (m^2 s^{-1})
Air	18.1×10^{-6}	1.20	15×10^{-6}
Water	1.00×10^{-3}	1.00×10^3	1.00×10^{-6}
Sea water	1.07×10^{-3}	1.02×10^3	1.05×10^{-6}

components: skin friction and pressure drag. Skin friction is a direct consequence of the viscosity of water and is greater the greater the surface area of the fish. Pressure drag is indirectly due to viscosity. Inertia tends to keep fluid particles moving in the direction they have already adopted, whereas frictional forces tend to align the fluid particles in conformity to the shape of the fish. At a critical point, which depends on the density and viscosity of the fluid together with the velocity and length of the fish, the inertial forces cause a separation of fluid from the fish. Consequently, the increase in the pressure of water at the front of the fish is not counterbalanced by an equal and opposite increase in pressure in the rear. This generates pressure drag. The relative importance of skin friction and pressure drag depends on the Reynolds number (Re). This is defined as:

$$Re = lU/v$$

where l is the effective length of the object in the fluid, U is velocity and v is kinematic viscosity. Re is the ratio of inertial to viscous forces. When the Reynolds number is low, the object is in a highly viscous environment, while at high Re, the environment is dominated by inertial forces (Table 1.2). Furthermore the value of Re is not just dependent on viscosity, but also on the size and velocity of the object.

Table 1.2 Effect of Reynolds number on relative importance of pressure drag—assuming that the object is a cylinder. (Data from Vogel 1988.)

Reynolds number Re	Pressure drag as % of total drag	Fish stage
10	57	Egg and larva
100	71	Larva
1000	87	Juvenile
10000	97	Juvenile/adult

At low Re, skin friction is of great importance. This depends on the area exposed to the friction and little on the orientation of the object. However, at moderate and high Re, inertial forces, and so pressure drag, are most important and depend on the orientation of the object. If the object is streamlined with a long, tapering tail, the tendency for fluid to separate from the object is reduced so that the pressure at the back is much closer to that at the front than for a non-streamlined object. A characteristic of many fishes is their streamlined appearance. Teleosts lay small eggs (chapter 6) so the eggs and larvae when they hatch emerge into a world of low Re dominated by viscous forces. Eggs and the larvae that are still dependent on yolk do not have a streamlined shape. However, as the larvae grow and as their muscles mature they become capable of swimming faster and so move into a range of Re values in which both frictional and inertial forces are important. With more growth, the young fish break out from this transitional environment into one dominated by inertial forces, where streamlining, that is an approach to the adult shape, will be advantageous (Webb and Weihs, 1986). Because of the effects of the density and viscosity on objects moving through water, the body shape of fishes is closely related to the mode of life (chapter 2).

The viscosity of water is strongly affected by temperature and to a much lesser extent by salinity (Figure 1.2). At 5°C, fresh water is 2.5 times as kinematically viscous as it is at 35°C. This temperature effect may be

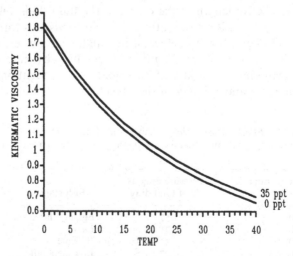

Figure 1.2 Effect of temperature (°C) on kinematic viscosity ($m^2 s^{-1} \times 10^6$).

important for small, larval and juvenile teleosts, which experience relatively high frictional drag.

For fish in open water, the effects of density and viscosity are experienced while swimming or holding station in a current. However, fishes like rays and flatfishes spend time on the substratum and the density of these fishes, although close to that of water, can be sufficent for the friction with the ground to allow the fish to hold station in a current without swimming (Webb, 1988). The orientation and movement of the fins and morphological devices that increase friction can also help the fish stay on the bottom. There is also the boundary layer caused by the viscosity of water, so near the substratum, current speeds are less than in the open water.

1.2.1.4 *Heat capacity.* Water has four times the heat capacity of air and because it is nearly 800 times more dense, it has a heat capacity per unit volume that is about 3000 times greater than air. Water can easily absorb the heat produced by metabolic processes in the fish and consequently, with a few exceptions, all fish are ectothermic (Graham, 1983). Their body temperature is close to that of the water surrounding them. The exceptions are large, pelagic fishes. They include the tunas (teleosts) and the makerel sharks (elasmobranchs), which maintain an elevated temperature throughout much of their body. The swordfishes and marlins (both teleosts) maintain an elevated temperature in the brain and adjacent tissues. This endothermy in fishes depends on a large body size and the presence of morphological adaptations that minimize heat transfer from muscles to the external environment. The endothermic fishes swim continuously, so heat is continually generated at a relatively high rate in the swimming muscles (Stevens and Neill, 1978).

For most fishes, because they are ectothermic, water temperature is the major, abiotic environmental identity—the master ecological factor (Brett, 1971).

1.2.2 *Solubility of gases and inorganic ions*

1.2.2.1 *Gases.* Fishes are essentially aerobic organisms. They require a supply of oxygen and produce carbon dioxide as one of the by-products of respiration. As they frequently use protein as a substrate for respiration, they also produce ammonia as a by-product. The solubility of gases in water is affected by temperature, salinity and pressure. Of these, temperature is the most important.

Oxygen dissolves poorly in water. A litre of water contains only about a thirtieth of the oxygen contained in a litre of air. The viscosity and density of water also mean that it requires more work to move water over the respiratory surface than the same volume of air. Solubility decreases with increases in both temperature and salinity (Figure 1.3). It also decreases with a decrease in pressure. Streams and lakes at high altitudes have a lower concentration of oxygen than those at low altitudes, other things being equal (Figure 1.4).

Both carbon dioxide and ammonia are much more soluble than oxygen in water. Consequently, it is the supply of oxygen rather than the removal of the metabolic waste products that can pose problems for fishes.

1.2.2.2 *Inorganic ions.* The average salinity of sea water is about 35 ppt, with the usual range in seas of 32–38 ppt. This gives an osmotic concentration of about $1000\,mosm\,l^{-1}$, approximately three times the osmotic concentration of the body fluids of teleosts. Sodium and chloride

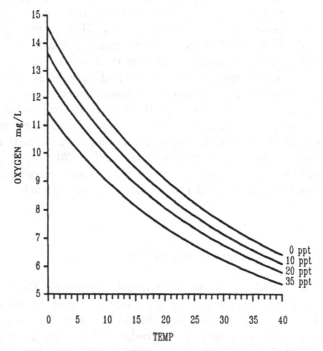

Figure 1.3 Effect of temperature (°C) and salinity on saturated oxygen concentration in water.

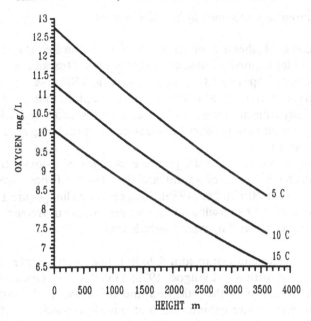

Figure 1.4   Effect of altitude and temperature on saturated oxygen concentration in water. C = temperature in °C.

ions are the major constituents of sea water (Table 1.3). Fresh water has a negligible osmotic concentration with bicarbonate often the most important ion (Table 1.3). Whereas the composition of sea water is relatively constant, that of inland waters depends on local conditions including the composition of local rocks and soils and the patterns of flooding and evaporation. Inland waters subject to evaporative loss or fed by saline springs can reach salinities far in excess of sea water.

Table 1.3   Concentration (mmol kg^{-1}) of major inorganic ions in average sea and fresh waters

Ion	Sea water	Fresh water
Chloride	548.3	0.23
Sodium	470.2	0.39
Sulphate	28.25	0.21
Magnesium	53.57	0.21
Calcium	10.23	0.52
Potassium	9.96	0.04
Bicarbonate	2.34	1.11

1.2.3 *Transmission of sound, light and electricity*

1.2.3.1 *Sound.* Fishes are sensitive to sound transmitted through water (Hawkins, 1986). Compared with air, water is a good transmitter of sound. In sea water at 35 ppt and 13°C, sound travels at $1500\,m\,sec^{-1}$ compared with $340\,m\,sec^{-1}$ in air. It also attenuates less rapidly than in air and so can potentially transmit information over long distances. Indeed, the sea is a noisy place and many fishes have devices for producing sound or use sound to detect prey.

Disturbances associated with pressure changes of a lower frequency than those characteristic of sound are detected by fishes through the lateral-line system (Bleckmann, 1986). Such pressure changes are generated by movement of the fish itself as it accelerates, by the movement of other objects in the water and by water turbulence.

1.2.3.2 *Light.* In contrast to sound, light is poorly transmitted by water when compared with air (Lythgoe, 1979; Douglas and Djamgoz, 1990). Because of the effects of the reflection and refraction at the surface, the light entering the water column has a strong downwards directionality. Once in the column, light is absorbed and scattered so it is unlikely that even the largest object can be seen at a distance much greater than 40 m in water. Even in the clearest oceanic water, the last traces of sunlight only penetrate to 1 km, while in turbid waters all the light has been absorbed within a few metres of the surface. The scattering and absorption of light is increased by the presence of suspended and dissolved matter in the water column.

Water also has a filtering effect on light. In clear water, the red end of the spectrum is absorbed first while the blue end of the spectrum penetrates much more deeply. However, in water that contains high loads of dissolved and suspended matter, the blue end of the spectrum is absorbed before the red end. Thus the spectral properties of light in water depend on depth and water quality. The levels of dissolved and suspended materials are highest in many fresh and coastal waters and these typically have a yellow or greenish tinge, whereas the unproductive, open oceanic waters are blueish.

Despite the poor transmission of light in water, many fishes have well-developed eyes. The visual pigments of many teleosts suggest that they have colour vision, although in most cases the evidence for this is still lacking. Elasmobranchs probably lack colour vision. The wavelengths of light allows it to transmit details of objects at high resolution, so even though it is transmitted only over short distances in water, it can provide fish with detailed information about objects within those distances.

1.2.3.3 *Electricity.* Animals generate electrical fields as an incidental by-product of neural and muscular activities. Some fishes can detect these fields and some even have special electric organs to produce characteristic electrical fields (Bleckmann, 1986). Salt water has a low electrical resistance caused by the presence of many inorganic ions. Some marines fishes can, however, find buried prey by detecting the electrical field generated by the hidden prey. Fresh water has a much higher resistance. The greatest development of the use of electrical fields for detection of prey, mates, and structures in the environment and even inter-fish communication has taken place in some freshwater families of teleosts. These fishes typically live in water that is turbid or they are nocturnal.

1.3 Diversity of fishes

Fishes represent just over 50% of all vertebrate species. Of the nearly 22 000 species of fishes recognized, 0.3% are agnathans, 3.7% are chondrichthyians and 96% are teleostomes (Nelson, 1984). There are more than 20 000 species of teleosts and they comprise 48.2% of all known vertebrates. Although fresh water covers only about 1% of the Earth's surface and accounts for less than 0.01% of its water, about 40% of all fishes live in fresh water. In the seas, 77.5% of marine species live in coastal and littoral waters. The upper 200 m of the open oceans, the epipelagic zone, is home to 1.9% of marine species. Oceanic waters deeper than 200 m, which by volume form by far the greatest component of the Earth's aquatic environment, are inhabited by the rest. Crude estimates suggest that marine fishes taken as a whole have 10 to 10 000 times more space available per individual than freshwater fishes (Horn, 1972). This reflects the relative emptiness of the deep oceans.

Species richness, that is the number of species found in an area, is greatest in shallow, tropical waters and is particularly high on coral reefs. It is also high in large tropical lakes and rivers. In both fresh waters and in the sea, species richness declines with increasing latitude. It also declines with increasing water depth, and for freshwater species with increasing altitude. The patterns of fish diversity suggest the following generalizations. There are more species in habitats that are structurally complex, such as coral and rocky reefs. For a given degree of structural complexity, the

number of species increases with the area of the habitat. Other things
being equal, species richness increases with the productivity of the habitat
unless the high level of productivity causes unfavourable abiotic conditions
to develop (chapter 2). Species richness decreases as the abiotic conditions
become less favourable for life, either because conditions become more
variable or because the conditions approach those at which life cannot
exist (chapter 2).

A brief description of the three classes of fishes will provide a background
for the succeeding chapters (see also Bond, 1979; Bone and Marshall, 1982;
Nelson, 1984; Moyle and Cech, 1988). A general classification of fishes is
given in the Appendix.

1.3.1 The Agnatha

As their name indicates, the agnathans are jawless fishes. Although they
have a long evolutionary history and were an important component of
the vertebrate fauna in the late Palaeozoic some 300–400 millions years
BP, today they are represented by only two orders, the hagfishes and the
lampreys (Figure 1.5). These modern forms have cartilaginous skeletons,
an unrestricted notochord and lack paired fins. The inner ear has only
two semi-circular canals compared with three in all the jawed fishes.
Although they lack the diversity of forms and modes of life found in the

Figure 1.5 Living agnathans: above—lamprey (*Petromyzon*) attacking a salmon; below—
hagfish (*Myxine*) scavenging on a dead fish.

jawed fishes, they are successful within the limited scope imposed by their morphological and physiological characteristics (Hardisty, 1979).

1.3.1.1 *Myxiniformes: the hagfishes.*   The eel-shaped hagfish are found only in the sea, indeed their body fluids are virtually isotonic with sea water. There are 30 to 35 species. Usually they live at depths between 25 and 600 m on bottoms of mud, silt or clay into which they will burrow. They have the ability to produce copious quantities of mucus (their name is derived from the Greek word for slime), which may deter predators. They have an endearing habit of tying themselves in a knot, which passes down the body clearing off excess mucus. The knot may also provide extra leverage when they are feeding. They are predators of invertebrates and scavengers on dead or moribund fish. Lacking biting jaws, they feed by the rapid eversion and retraction of horny teeth situated on the sides of the mouth. Their eyes are reduced, but they have three pairs of barbels surrounding the terminal nasal opening and the mouth and presumably locate their food by olfaction and taste. The young develop directly from large eggs.

1.3.1.2 *Petromyzontiformes: the lampreys.*   There are about forty species of lamprey, including four species that live in the Southern Hemisphere. They live in temperate regions and unlike the hagfish are found in both fresh and sea water. The life history has a distinct larval stage, the ammocoete, which metamorphoses into the adult form. The ammocoete is a small filter feeder that lives buried in the bottom of rivers and streams. After metamorphosis, some lampreys become sexually mature in the river and do so without feeding as adults. These are small species (< 200 mm total length). Other species migrate either to a lake or to the sea where they feed on fishes, growing to a larger size (300–800 mm). For spawning, they migrate back into rivers, ceasing to feed then dying after spawning.

The adult lamprey feeds as an ectoparasite or predator. If the victim survives, it bears the disc-shaped scar left by the lamprey's suctorial mouth (see chapter 5). Lampreys can be a major pest of commercial or recreational fisheries (see chapter 7).

1.3.2 *Chondrichthyes*

The cartilaginous fishes are represented by about 800 species. Within a restricted adaptive radiation they form a distinctive, and for a few species,

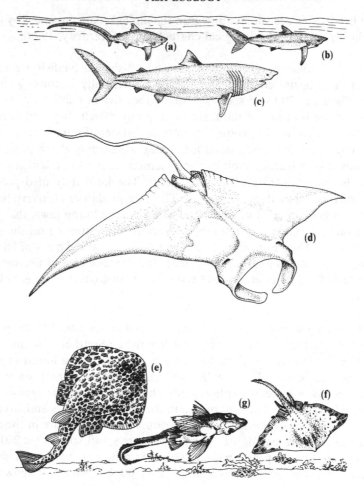

Figure 1.6 Living chondrichthyians with typical adult sizes. Pelagic forms: (a) thresher shark (*Alopias*) at 4 m; (b) blue shark (*Prionace*) at 3.5 m; (c) basking shark (*Cetorhinus*) at 7 m; (d) manta ray (*Manta*) at 6 m. Benthic forms: (e) electric ray (*Torpedo*) at 1 m; (f) thornback ray (*Raja*) at 1 m; (g) chimaera (*Harriotta*) at 1 m.

a feared component of the fish fauna (Figure 1.6). Although predominantly a marine group, both the sharks and, to a greater extent, the rays have freshwater representatives. These are restricted to tropical and sub-tropical waters. The Chondrichthyes are divided into two distinctive sub-classes that have had long, separate evolutionary histories although they share several chondrichthyian characteristics. They lack a swim-bladder or lung, so they tend to sink in the water column unless by swimming they generate hydrodynamic lift. All are carnivores and the main difference is between those forms that live in the water column and those that take advantage of their lack of buoyancy by spending much of their life on or close to the bottom.

1.3.2.1 *Chimaeriformes.* These little-known, peculiar fish belong to the sub-class Holocephali. There are about thirty species and because of their appearance they have acquired a variety of names including chimaera, ratfish, rabbitfish and spookfish. They are all marine and usually found in deeper waters. As adults, they are between 600 mm and 2 m in length. They have large eyes, large pectoral fins and a slender tail. The gills are covered by a fleshy operculum. The well-developed pavement-like teeth account for some of their common names. In contrast to the other chondrichthyians, tooth replacement in the chimaeras is slow. They feed on invertebrates and other fish. As with all chondrichthyians, fertilization is internal. The male has claspers on the head and the abdomen, which are presumably used to grasp the female during mating.

1.3.2.2 *Elasmobranchii.* This sub-class contains the more familar sharks and rays. They have separate gill slits and a spiracle, which, in the bottom-living forms, allows the entry of water to ventilate the gills. All elasmobranchs are carnivores, some like the pelagic sharks actively chasing their prey. Others rely on stealth and camouflage to ambush the prey. The two largest sharks, the basking and the whale shark, filter feed on zooplankton. The main division is between the sharks (Selachimorpha) and the skates and rays (Batidoidimorpha). This is basically a division between torpedo-shaped, streamlined fishes (sharks) and dorsoventrally flattened fishes in which the pectoral fins are greatly expanded into 'wings', which attach to the side of the head (skates and rays). Although most of the skates and rays are bottom-dwellers, some—the eagle rays and mantas—have adopted a secondary life in the water column using their pectorals as 'wings' to generate lift (Figure 1.6).

Within the elasmobranchs, several unusual adaptations have evolved. The thresher shark has a greatly elongated upper lobe to its tail, which is used to thrash the water, concentrating a shoal of fish making it easier to attack. Some rays have a formidable venomous spine and others have electric organs on the head, which can deliver a powerful electric shock. The saw shark and the saw fish have an elongated rostrum armed with lateral teeth that is used to slash at fish.

1.3.3   Osteichthyes (Teleostomi)

These fishes have a skeleton that is, at least partly, formed of true bone. They also have a swim-bladder or lung, which, because it contains gases including nitrogen and oxygen, reduces the overall density of the body. This can give the fish neutral buoyancy, so it does not need to swim to hold its position in the water column. The gills are protected by a bony operculum. Four evolutionary lineages of bony fishes can be recognized, but the relationships between the four are uncertain. Three of the four have few living representatives. The fourth, the Actinopterygii, includes the teleosts, the most successful of all the fishes.

1.3.3.1   *Dipneusti, Crossopterygii and Brachiopterygii.*   These all have fins with a fleshy lobe, although in some forms the fin has become greatly modified.

The Dipneusti or lungfishes are found in Australia, Africa and South America in swamps and waters likely to become de-oxygenated. They have functional lungs, although the Australian representative, *Neoceratodus*, uses primarily gill respiration. The African (*Lepidosiren*) and South American (*Protopterus*) forms can survive out of water using lung respiration and can survive periods of drought by aestivating in burrows. They are elongated, large fish, reaching lengths of up to 2 m and are either omnivorous or carnivorous.

The Crossopterygii has one living representative, the famous living fossil found in the Indian Ocean, the coelocanth, *Latimeria* (Balon *et al.*, 1988). This is also a carnivore, feeding on fish and squid. It lives at depths of 70–600 m.

The Brachiopterygii is represented by the single order, Polypteriformes, containing ten species of *Polypterus*. These elongated, predatory fishes live in swamps in Africa and can respire using their lungs. They have larvae, which, like the larvae of the African and South American lungfishes, have external gills and so superficially resemble the larvae of amphibians.

1.3.3.2   *Actinopterygii.*   The most primitive ray-finned fishes belong to the infraclass Chondrostei, whose only living representatives are the sturgeons and paddle fish—the Acipenseriformes. These fish retain a spiracle and the tail is asymmetrical with the upper lobe larger than the lower (heterocercal condition), as it is in the Chondrichthyes. The sturgeons are of some economic importance because the eggs of a few species are eaten as caviare, a highly prized delicacy. They are confined to the holarctic

Northern hemisphere. Some species live their entire life in fresh water, others spawn in fresh water but migrate to estuaries or coastal sea waters to feed and grow (see chapter 4). They are large slow-growing fish, only attaining sexual maturity after about 10 years. They have barbels on the underside of the head and a protrusible mouth. They feed on invertebrates living in or on the bottom of the sea or river. Paddle fish are found only in the Mississippi River system in North America and Yangstze River in China. They have a long snout with tiny barbels and lack the bony scutes on the flanks carried by the sturgeons. The American paddle fish feeds by filtering out zooplankton.

The infraclass, Neopterygii, can be conveniently divided into the holosteans, with few species and the highly diverse teleosts. The holosteans, which are evolutionary more primitive than the teleosts, have lost the spiracle. The tail fin is superficially symmetrical but its skeleton still shows traces of the asymmetrical, heterocercal condition. The Lepisosteiformes or gars are elongated, predatory fish found in fresh and brackish water in eastern North America. They live in quiet, weedy water and can supplement gill respiration with the use of atmospheric oxygen. The Amiiformes has only a single representative, the bowfin, *Amia*. This is confined to fresh waters in eastern USA. It is also a predator found in warm shallow waters and can use atmospheric oxygen for respiration.

The teleosts have a fully symmetrical tail (homocercal condition). The gas-bladder is used primarily as a hydrostatic organ providing buoyancy, and only in a few species functions secondarily as a lung. The bones of the skull show considerable modification and these modifications form the basis of the extensive adaptive radiation in modes of feeding shown by the teleosts (chapter 5). The scales are reduced compared with those of the other bony fishes. The characteristic ridge and valley structure of these scales assumes importance in the study of the ecology of teleosts because changes in the pattern provide a means of ageing the fish (chapter 5).

Teleosts are found in both fresh and sea water, from close to the zone of permanent snow at high altitudes to the abyssal depths of the oceans. The most diverse teleost faunas, in terms of numbers of species, occur in warm, shallow, tropical and sub-tropical seas and in the catchments of large tropical rivers. However there are species living in the Arctic and Antarctic seas where the water temperature is permanently below 0°C.

Primary seawater teleosts, which include most marine fishes, have evolved in the sea and are intolerant of water of reduced salinity. Primary freshwater teleosts, of which the cypriniformes and characiformes are the most prominent, have a long evolutionary history in lakes and rivers and usually have a low tolerance of salt water. Secondary freshwater teleosts, such as the Cichlidae, have close relatives that are marine fishes and can tolerate brackish or salt water. Diadromous teleosts, like many salmonids and some eels, spend part of their life in fresh water and part in the sea and can tolerate a wide range of salinities (chapter 4).

The diversity of the teleosts precludes a detailed description of all the orders in this Introduction. Examples that will be met repeatedly in the text are worth introducing at this point. The descriptions follow, approximately, a sequence from primitive to evolutionarily advanced teleosts. This progression is marked by several changes in morphology. The pelvic fins migrate anteriorly until they come to lie under or even in front of the pectoral fins. The latter tend to migrate dorsally to lie on the side of the body rather than ventrally. In the more primitive teleosts, the fins are supported only by soft rays, but in the more advanced teleosts some fins contain spiny rays. The swim-bladder, in the primitive condition, is connected to the alimentary canal (physostome condition), but this connection is lost in the physoclistous teleosts.

The Osteoglossiformes include the largest freshwater teleost, *Arapaima gigas*, of the Amazon basin. It reaches a length of 2–3 m. The order also includes two families, the Mormyridae and Gymnarchidae, which have highly developed electric organs. They live in fresh waters in Africa, often in turbid water or they are active at night. The electric organs are used both for the detection of objects in the water and for communication between individuals.

The Anguilliformes are the true eels. They are mainly a marine group, but a few species are diadromous including the european eel, *Anguilla anguilla*.

The Clupeiformes are the herrings, spratts and anchovies. They are streamlined, pelagic fishes feeding on plankton. Although mostly marine, there are many freshwater clupeoids in lakes and large rivers. They usually live in large shoals. Clupeoids support many large commercial fisheries throughout the world (chapter 7).

The super-order Ostariophysi includes three orders, the Characiformes (characins), Cypriniformes (cyprinids) and the Siluriformes (catfish), which dominate many tropical and temperate freshwater habitats. With the exception of one family, the sea catfish (Ariidae), they are confined to fresh or slightly brackish waters. The Ostariophysi are characterized by a row of small bones, the Weberian ossicles, which link the swim-bladder with the ear. This linkage enhances the sound reception of the ostariophysians and allows them to hear sound of higher frequencies than fish, which lack a connection between the swim-bladder and ear (Hawkins, 1986). The ostariophysians also produce an 'alarm substance' when the skin is damaged causing a fright response in the companions of the damaged fish. The characins are restricted to tropical fresh waters, and together with the silurids dominate the fresh waters of South America and are an important element of the African fauna. The cyprinids are absent from South America, but are important in the fish faunas of Africa, South East Asia, North America and Eurasia. They include such familiar European fishes as the minnow (*Phoxinus phoxinus*), roach (*Rutilus rutilus*), bream (*Abramis brama*) and tench (*Tinca tinca*). Cyprinids are important in aquaculture in China, India and other countries (chapter 7). The silurids occur on all three continents and although predominantly tropical, there are catfishes in temperate fresh waters in both North America and Eurasia.

The Salmoniformes include two families that have attracted much attention from fish biologists. The Esocidae is a holarctic, freshwater family of carnivores, particularly fish-eaters. The northern pike, *Esox lucius*, is a typical example. The Salmonidae is also a northern hemisphere family with a holarctic distribution. The family has both freshwater and diadromous species; indeed some species show both types of life history. The family includes the Pacific salmon (*Oncorhynchus* spp.), the Atlantic salmon and trout (*Salmo* spp.), the chars (*Salvelinus* spp.) and the white fish (*Coregonus* spp. and others). The family is economically important, supporting both commercial and recreational fisheries and with several species farmed in aquaculture facilities (chapter 7). The species are all carnivores, but the food can range from zooplankton (especially in the white fish) to other fish in the larger salmon and trout (chapter 5).

The Stomiiformes, Aulopiformes and Myctophiformes are deep-sea carnivores. Although they are probably important to the ecology of the deep seas, their habitat means that their biology is poorly known in comparison to shallow water fishes.

Of great commercial importance are the Gadiformes. With one exception, the burbot (*Lota lota*), they are all marine species. The order includes cod (*Gadus morhua*), pollack (*Pollachius pollachius*), haddock (*Melanogrammus aeglefinus*) and whiting (*Merlangius merlangus*), all of which support fisheries.

The Lophiiformes are marine fishes found in both shallow and deep seas. They are called angler fish for they attract their prey, usually other fish, close to their enormous mouth with a lure formed from a modified spine of the dorsal fin. In one family of deep-sea anglers, the males are parasitic on the females (chapter 6).

Of little importance for fisheries but an order that has attracted much study is the Cyprinodontiformes. This includes the flying fish (Exocoetidae) of the open seas as well as many small fishes found in fresh and brackish waters. Some members, unlike most teleosts, have internal fertilization and bear their young live. Even more unusually, in one family (Poeciliidae), there are parthenogenetic species (chapter 6). Many are popular aquarist species including the guppy (*Poecilia reticulata*) and the swordtails (*Xiphophorus*).

The Gasterosteiformes includes only about nine species confined to coastal seas and fresh waters of the northern hemisphere. However, the order includes the threespine (*Gasterosteus aculeatus*) and other sticklebacks, which have been the subjects of intensive behavioural, ecological, evolutionary and physiological studies. Possibly related are the Syngnathiformes. This entirely marine order includes strangely-shaped fishes—the pipefish, sea horses, cornetfish and trumpetfish—found in weed beds and on rocky and coral reefs.

Also usually found on rocky reefs are the Scorpaeniformes. These are mainly marine fishes, including the rockfish (Scorpaenidae). However, the bottom-living sculpins (Cottidae) are found in both sea and fresh water. In Lake Baikal (USSR), the cottids have evolved numerous species that are found only in this deep and ancient lake.

Nearly 8000 species are Perciformes. Although predominantly marine, there are many freshwater perciformes including the European and American perch (*Perca fluviatilis* and *P. flavescens*). The Chaetodontidae (butterflyfish), Pomacentridae (damselfish), Acanthuridae (surgeonfish) and Scaridae (parrotfish) are important in diverse fish faunas of coral reefs. Two families, the Labridae (wrasses) and Serranidae (sea basses), include hermaphroditic species (chapter 6). Two freshwater families of Perciformes have attracted much study. The Cichlidae are found in Africa and Central and South America and India. They include mouthbrooders—species in which the eggs and newly hatched young are brooded in their mother's mouth. In other cichlids, the eggs are laid on the substrate but both parents guard and tend them. In the Great Lakes of Africa, the cichlids have shown a wide adaptive radiation forming flocks of closely-related species and showing a profusion of feeding habits (Fryer and Iles, 1972) (chapter 5). A few cichlids are important in aquaculture in tropical and sub-tropical countries. In the lakes and rivers of eastern North America, the Centrarchidae (sunfish and basses) have shown a lesser but comparable adaptive radiation. Experimental and field studies on the centrarchids have made major contributions to the understanding of the community structure and feeding ecology of freshwater fishes (chapters 3 and 5).

The Pleuronectiformes are marine, bottom-dwelling species characterized by a metamorphosis that transforms a bilaterally symmetrical, pelagic larva into the laterally compressed flatfish lying on one side. During the metamorphosis one eye migrates to the upper side. Although the resulting shape is reminiscent of that of skates, the manner of achieving it is completely different. Many flatfishes such as the plaice (*Pleuronectes platessa*), halibut (*Hippoglossus hippoglossus*) and sole (*Solea solea*) are fished commercially.

The Tetraodontiformes includes some of the strangest-looking teleosts. All are marine and most live in shallow tropical or sub-tropical waters including coral reefs. The Balistidae are the trigger fish with a first dorsal spine that can be locked in an upright position. The Ostraciidae, boxfishes, are almost entirely encased in a carapace of thickened scale plates. Puffers have, as a defence response, the ability to inflate a sac-like extension of the gullet, expanding the fish almost like a balloon. The Molidae (sunfish) are large, lazy-swimming, pelagic fish that lack a tail stalk.

Aspects of the ecology of many of these fishes will be described in the following pages.

1.4 Relationships

Although the word ecology has attracted to it a variety of scientific, political and philosophical implications, a useful working definition is the scientific study of the interactions between organisms and their abiotic and biotic environments that determine the distribution and abundance of the organisms (modified from Krebs, 1985). The abiotic environment is provided by the physical and chemical conditions experienced, the biotic environment by the other organisms encountered.

At the heart of ecology is the study of the effects of the abiotic and biotic conditions on the reproductive success of individuals, that is the number of offspring that an individual produces in its lifetime. The abundance and distribution of a population will depend on the average success of individuals in that population in producing offspring taken together with the patterns of movement of the individuals. A population has continuity over time but through dispersion and migration (chapter 4), its spatial distribution may change. The population or a portion of the population may occupy one geographical location at one time but a quite different one at another time.

The number of species that co-exist in the same location, the fish assemblage, will depend on the reproductive success of the individuals of those species that do or could have access to the area. For some species, their reproductive success will be affected by interactions with other species of fish and other organisms (chapter 3). Those species that interact either directly or indirectly at a location form a community (Giller and Gee, 1987). It is often convenient to consider only those interacting species that have some taxonomic affiliation. For example, the fish community (in the wide sense of the term) of a lake is part of a wider community of interacting organisms that includes microorganisms, phyto- and zooplankton, macrovegetation, and other invertebrates and vertebrates.

To be reproductively successful, a fish requires the energy and materials that are provided in its food. Consequently, the presence of fishes affects the flux of energy and the cycling of elements such as carbon, nitrogen, and phosphorus. The analysis of energy fluxes and element cycling forms a major part of the study of the ecology of ecosystems, which emphasizes the interrelationships between communities of living organisms and the physical environment over a relatively wide geographical scale.

The study of ecology is hierarchical—comprising individuals, populations, communities and ecosystems—but at its core is the study of individual organisms attempting to be genetically represented in the next generation given the opportunities and hazards generated by the abiotic and biotic environments (Begon *et al.*, 1989).

The success of this attempt will depend on the individual's survival through to sexual maturity and its subsequent fecundity (number of eggs laid by female or fertilized by a male). In most fishes, both survival and fecundity are influenced by size, so the individual's growth rate will also be a factor. Growth, survival and fecundity are components of fitness that determine the relative success of an individual in being genetically represented in the next generation compared with other individuals in the population (Figure 1.7a). The abundance of the population will depend on the balance between the rate of gain (births and immigration) and rate of loss (deaths and emigration (Figure 1.7b)).

The reproductive success of individuals in a population will depend on processes that operate on different spatial and temporal scales. An individual may be influenced by events that occurred many kilometres away and many months before. An example is the effect of the major climatic and oceanographic event known as the El Nino Southern Oscillation (ENSO) (Philander, 1990). This affects much of the Pacific Ocean and adjacent lands. The consequences include a change in the

(a)

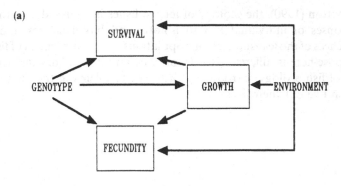

(b)

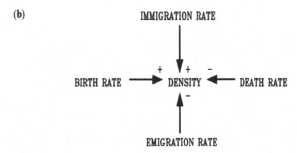

Figure 1.7 Factors affecting fish populations: (a) interrelationships between components of fitness, genotype and environment; (b) factors affecting density of a fish population.

temperature and nutrient content of the waters off the Peruvian coast. These changes cause a decline in the phyto- and zooplankton populations, which are the food of the Peruvian anchoveta (*Engraulis ringens*). The average reproductive success of individual anchovetas present during an ENSO event declines and population abundance falls with consequences for the birds and fisherman dependent on the anchovetas (chapter 7). The immediate cause of the decline in the reproductive success of the anchovetas is the lack of food, perhaps combined with relatively high sea temperatures. However, the chain of events that produced these conditions takes place over spatial scales of the order of several thousand kilometres and time-scales of months and years. The survival of a larval anchoveta may depend on the food particles it encounters in a few litres of water over a few hours. Yet the events that help to determine the concentration of those particles are on an oceanic scale.

In Wootton (1990), the ecology of teleost fishes is analysed in terms of the responses of individual fish to abiotic and biotic factors and the consequences of those responses for populations and communities of fishes. The purpose here is different. It is to provide an introductory descriptive account of fish ecology, introducing the main problems in the context of where fish live out their lives.

CHAPTER TWO

EFFECTS OF ABIOTIC ENVIRONMENTAL IDENTITIES ON DISTRIBUTION

2.1 Introduction

Fish species have restricted geographical distributions. For some species, this distribution may cover a vast geographical area. Many species of tuna have a transoceanic distribution. Other species are found only in tiny areas. The Devil's Hole pupfish, *Cyprinodon diabolis*, occurs only in a limestone sink in Nevada, USA (Naiman and Soltz, 1981). The fundamental cause of these restricted distributions is the heterogeneity of the physical environment in which the evolution of the fishes has taken place. At an evolutionary level, this heterogeneity is important because physical barriers will prevent gene flow between populations and so allow speciation (Mayr, 1963). At a biogeographical level, physical barriers can prevent a species that was evolved in one locality from colonizing another area to which it is physiologically well-adapted.

The barrier may be something as simple as an impassable waterfall, or it may be a stretch of water whose properties the species cannot physiologically tolerate. Sea water can act as an effective barrier to primary freshwater fishes because of their inability to osmoregulate in high salinities. The high proportion of freshwater fish species (chapter 1) probably reflects both evolutionary and biogeographical effects on species richness. Freshwater habitats are much more broken up than the continuous stretches of the oceans. In fresh waters, populations are more likely to be isolated from gene flow from other populations and so more likely to speciate. There are also likely to be both land and sea barriers preventing dispersal. Evolutionary and biogeographical aspects of species distribution, although important, will not be explicitly considered further, rather attention will be focused on the contemporary role of abiotic identities in determining species distributions.

The importance of the various abiotic identities on fish distribution varies with the nature of the aquatic environment. Identities that may play a dominant role in upland streams will be of little or no importance at great depths in the oceans. For fishes, the dominant identities are often temperature, oxygen, salinity and water movement. Other identities such as pH, the presence of toxic substances and pressure may be important in particular environments.

Fishes can only survive within a certain range of an abiotic identity such as temperature. Outside that range, the fish dies—the identity acts as a lethal factor (Fry, 1971). Temperature, oxygen and salinity can all act as lethal factors. Some species can tolerate a wide range of an identity. This ability is described by the prefix eury-, so that species tolerant of a wide temperature range are eurythermal and those tolerant of a wide salinity range are euryhaline. Species that can tolerate only a narrow range of an identity are described by the prefix steno-, so that species may be stenothermal or stenohaline. The prefix meso- is sometimes used to indicate intermediate tolerance (Hokanson, 1977). Table 2.1 lists some characteristics of steno-, meso- and eurythermal freshwater fishes from northern latitudes.

The zone of tolerance describes the range of an identity over which survival is possible, but active feeding, growth and reproduction may take place only within narrower ranges (Figure 2.1). The ranges of these identities within which a species can successfully maintain itself by natural recruitment partly define the fundamental niche of the species (Giller, 1984). (The other component of the fundamental niche is diet as discussed in chapter 5.) The overall geographical distribution of a species will largely reflect its fundamental niche; however, interactions with other organisms may restrict the species to a range less than that predicted from its

Table 2.1 Characteristics of steno-, meso- and eurythermal northern, temperate, freshwater fishes (based on Hokanson, 1977 and Varley, 1967)

Criteria	Stenothermal[a]	Mesothermal[b]	Eurythermal[c]
Gonadal growth phase	Summer at $<20°C$	Autumn and winter at $<12°C$	Spring and early summer at $>12°C$
Time of spawning	Autumn to spring at $<15°C$	Spring at $3–23°C$	Spring to autumn at $15–32°C$
Physiological optimum	$7–20°C$	$14–28°C$	$20–30°C$
Upper lethal temperature	$<26°C$	$28–34°C$	$>34°C$

[a] e.g. Salvelinus fontinalis, Salmo trutta, Oncorhynchus mykiss, Coregonus artedii, Cottus gobio.
[b] e.g. Catostomus commersoni, Esox lucius, Perca spp.
[c] e.g. Micropterus salmoides, Lepomis spp., Tinca tinca, Abramis brama.

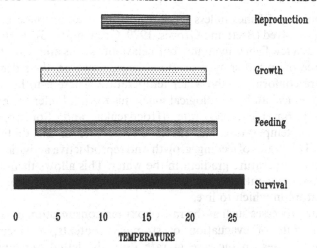

Figure 2.1   Approximate temperature ranges (°C) for the life processes for the three-spined stickleback (*Gasterosteus aculeatus*) from mid-Wales.

fundamental niche. These biotic interactions are discussed in chapter 3. A species may temporarily occupy an area, but because of the range of environmental conditions to which that area is subject, the species cannot successfully complete its life history there. Either, the species migrates to areas that allow it to complete its life-history (chapter 4), or the species dies out and its presence in that area depends on episodes of colonization from more favourable environments. The environments accessible to a species must provide conditions that allow some individuals to successfully complete all stages of their life-history so that new generations are produced. In other words, the environments must provide ontogenetic continuity. This requirement is crucial in the rehabilitation of habitats that have been degraded by man's activities (chapter 7).

The tolerances of individuals may change during their ontogeny. This is seen most dramatically in species that migrate between fresh and sea water (chapter 4). Tolerances may also change seasonally.

2.2   Effects of abiotic identities

2.2.1   *Temperature*

Temperature controls the maximum rate at which chemical reactions can occur. Changes in temperature will have direct effects on all aspects of the metabolism of fishes unless mechanisms for evading these pervasive effects have evolved (Brett and Groves, 1979; Graham, 1983). As described in chapter 1, a few fishes, including both elasmobranchs and teleosts, show some degree of endothermy. For most fishes, however, their deep body temperature conforms to the water temperature. These ectotherms have some biochemical and physiological mechanisms that buffer the effects of temperature on metabolic rate (Hochachka and Somero, 1984). Nevertheless, temperature has effects on the rate of metabolism and consequently on rates of feeding, growth and reproductive activities. Fishes can detect a temperature gradient in the water. This allows them to exert some behavioural control over their body temperature by selecting a range of temperature in which to live.

For some processes such as the rate of oxygen consumption by a resting fish or the rate of evacuation of stomach contents, an increase in temperature causes an increase in rate until the lethal temperature is reached (Figure 2.2a). For other processes, including the rate of food consumption and growth rate, there is an optimum temperature at which the rate is a maximum. At temperatures lower or higher than the optimum, the rate declines (Figure 2.2b). Different processes have different optimum temperatures (Elliott, 1981).

2.2.2 Oxygen content

The rate at which oxygen can be supplied to the respiring tissues will limit the rate of aerobic metabolism (Pauly, 1981). As a fish increases in size its demand for oxygen also increases. However, the rate of respiration is not directly proportional (isometric) to body weight, rather the relationship is allometric. This means that the rate of oxygen consumption per unit mass decreases as the fish gets heavier (Figure 2.3).

Although oxygen is relatively insoluble in water, the gills and related adaptations for respiration in fishes allow effective functioning when the water is saturated with oxygen. As the oxygen level drops below saturation, a critical oxygen concentration is reached below which metabolic rate and other processes are limited by oxygen supply. Eventually, an oxygen concentration is reached that the fish cannot survive indefinitely—the lethal level. Because the solubility of oxygen decreases with an increase in temperature (Figure 1.3), while the metabolic rate of fishes increases with an increase in temperature (Figure 2.2), the danger of reaching a lethal level of oxygen is usually higher in warm than in cold waters unless some barrier prevents diffusion of oxygen from the air into the cold water.

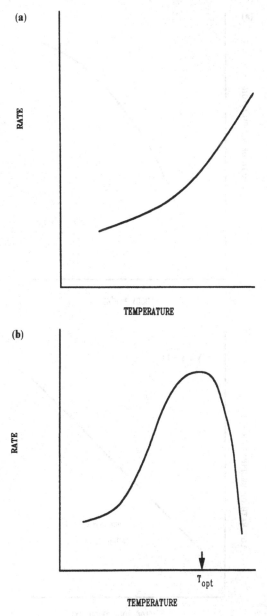

Figure 2.2 Schematic diagrams of the effect of temperature on rate of biological processes: (a) rate increases over zone of temperature tolerance; (b) rate has a maximum at the optimum temperature given by T_{opt}.

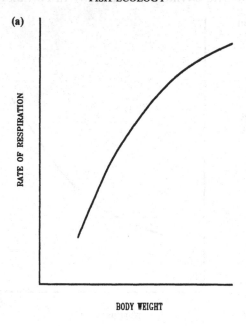

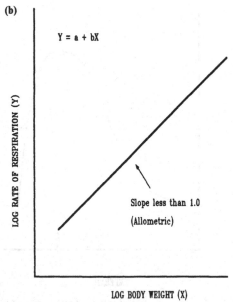

Figure 2.3   Schematic diagram of effect of body weight on rate of oxygen consumption: (a) both axes in arithmetic units; (b) both axes in logarithmic units.

Low oxygen concentrations in water are often caused by the presence of decaying, organic matter. These circumstances also tend to generate toxic gases such as hydrogen sulphide and methane.

The tolerance to low oxygen concentrations varies (Varley, 1967). Stenoxic species like trout and salmon require well-oxygenated waters containing more than $10\,mg\ O_2$ per litre. Mesoxic species, including pike and perch, require a minimum of $5\,mg\ 1^{-1}$. Euryoxic species, which include cyprinids like the carp and tench, can tolerate oxygen concentrations as low as $0.5\,mg\ 1^{-1}$.

2.2.3   Salinity

Most fishes spend all their lives in water that changes little in its salinity. However, some live in environments such as estuaries where salinity can change rapidly. Other species move, during migration, between waters that differ greatly in their salinity (chapter 4). These species have to meet any additional energy costs of osmotic and ionic regulations caused by

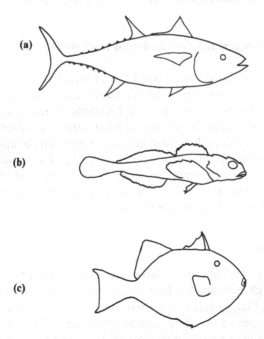

Figure 2.4   Fish body shapes adapted for: (a) fast cruising, *e.g.* tuna; (b) rapid acceleration and turning, *e.g.* cottid; (c) manoeuvrability *e.g.* trigger fish.

the changes in salinity. Occasionally, sudden changes in salinity may cause catastrophic mortalities as, for example, when the sea breaches a barrier and floods into a freshwater lagoon. Some fish follow gradients in salinity to find their way between waters that differ in salinity (chapter 4).

2.2.4 *Water movement*

Because of its density, moving water has a high momentum. This poses problems for any organisms trying to occupy habitats such as rivers and the littoral zones of lakes and seas where there is much turbulence. Body shape is an important trait because it determines the pattern of water flow past the fish, whether the fish is simply holding station in a current or is moving (chapter 1). Body form also affects characteristics such as the ability to manoeuvre effectively, accelerate rapidly, cruise long distances or to live close to the substratum (Figure 2.4). A body shape that makes a fish highly manoeuvrable is incompatible with sustained cruising or rapid acceleration and turning (Webb, 1984). Consequently, the shape of a fish often strongly reflects its typical way of life.

2.3   Abiotic factors and the distribution of river fishes

Stream order is a useful classification of river structure when describing the distribution of fishes. First-order streams are those that have no tributaries. The junction of two first-order streams creates a second-order stream, and the junction of two second-order streams, a third-order stream and so on. A convenient qualitative distinction is between the upper regions of the river (lower-order streams), the rhithron, and the lower regions (higher-order streams), the potamon. These regions differ in the physical characteristics of the river course and consequently in the relative importance of abiotic identities for fishes.

2.3.1 *The rhithron*

The upper sections of rivers are usually characterized by a relatively steep gradient and a high and variable discharge of water. The shallow, turbulent waters are highly oxygenated, but their small volume means that their temperature changes as the air temperature changes. Even in temperate south-west England, a daily fluctuation of 9.5°C has been recorded in a low-order stream (Webb and Walling, 1986). The typical morphology of upland rivers is an alternation of riffles and pools. The riffles are shallow stretches where water velocities are high and flow turbulent. In pools, the water is deeper so the flow is slower and less turbulent. In some Caucasian streams, the average flow rate measured at low water was $1.25\,\mathrm{m\,s^{-1}}$ through the riffles but only $0.6\,\mathrm{m\,s^{-1}}$ through the pools (Welcomme, 1985). The velocity of the water determines the sediment load carried. Finer sediment tends to be deposited in the pools, while coarser particles form the bed of riffles. As the rhithron can experience high rates of water flow at times of high rainfall or rapid snow-melt, the bed is often unstable with sediment washed downstream in the flood waters. Such rivers often cut deeply notched valleys, so even though water levels can rise quickly, the flow is contained within the valley and only a restricted area is flooded.

Fish in this environment must be adapted to cope with the variable water flows and their consequences and often daily and seasonal fluctuations in temperatures. In dry seasons, the stream may break up into a series of pools restricting the living space available to fish. In temperate and boreal zones, the build-up of ice during the winter may also restrict the living space. At high altitudes, a further possible hazard is ultraviolet (UV) light, which will cause skin-burning in fish exposed in shallow waters.

Some fishes in the rhithron have a body shape that allows them to hold station in rapidly moving water. Salmonids like the brown trout (*Salmo trutta*) of northern European streams provide an example with their stream-lined bodies (Figure 2.5), although the trout take advantage of slack water behind obstructions, moving into the stream flow to collect food. Other species have morphological adaptations for living on the steam bed, using crevices and holes between stones where current speeds are lower. A European example is the bullhead (*Cottus gobio*), which lacks a swim-bladder and has a heavy head. The bullhead has a body form that allows it to accelerate rapidly over short distances. Morphological traits can increase the friction with the bed (Welcomme, 1985; Webb, 1988).

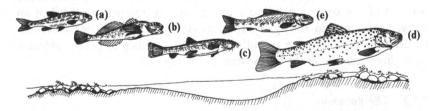

Figure 2.5 Representative fishes of the rhithron fish assemblage from western Europe: (a) minnow (*Phoxinus phoxinus*) (80 mm); (b) bullhead (*Cottus gobio*) (120 mm); (c) stone loach (*Noemacheilus barbatulus*) (100 mm); (d) brown trout (*Salmo trutta*) (200 mm); (e) juvenile salmon (*S. salar*) (120 mm).

Modified paired fins or the mouth are used as a sucker. Spines may also act as anchors. Bottom-living species typically are streamlined with a slightly humped back and a flattened ventral surface, which set up hydrodynamic forces that tend to force the fish onto the bottom (Welcomme, 1985). These traits can be extremely effective. When several species were tested for their ability to maintain their position in a current, the species that performed best was a bottom-living catfish, *Plecostomus*. It did this not by resisting the flow by swimming, but by attaching to the bottom with its sucking disc (Fricke *et al.*, 1987). Some species, for example the South American catfish *Astroblepus*, that live in torrential waters can ascend vertical surfaces using their suctorial apparatus (Norman, 1963).

Differences in body shape and other morphological traits will allow species to occupy microhabitats within the stream that differ in depth and flow characteristics. This microhabitat segregation may be important in minimizing inter-specific competition between species that have similar diets (Wikramanayake, 1990).

In cool temperate and boreal zones, fishes of the rhithron like the salmonids are stenothermal, preferring cool ($< 20°C$) waters. They are also intolerant of low oxygen conditions. These traits reflect the characteristics of the cool, well-oxygenated headwaters of many river systems. Nevertheless, in other geographical areas, the fish fauna of low order streams can experience harsh abiotic conditions. In the southern Great Plains of USA, daily temperature ranges of $10–13°C$ are common in summer. During dry periods, prairie streams dry up leaving a series of disconnected pools. In these pools and backwaters, the oxygen concentration can drop to levels below 1 ppm. The stream fishes living in these environments must be eurythermal and euryoxic.

A comparison of the temperature and oxygen tolerances of species of the cyprinid *Notropis* from prairie and more benign upland streams suggested that the species found in the prairie streams were both more tolerant of low oxygen conditions and more selective when placed in an oxygen gradient than the upland species (Matthews, 1987). The fishes most successful at recolonizing recently dried-out stretches of prairie streams were those most tolerant of low oxygen conditions.

2.3.2 *The potamon*

In this zone, the gradient of the river slackens and the morphology of the river course becomes more complex. In addition to the main channel of the river, there is often a flood plain, which is inundated seasonally. The erosion and deposition patterns of the river create backwaters, oxbow lakes, levees and swamps. Thus in the floodplain, there are aquatic habitats that experience little or no water flow when the river is confined to its main channel. In these regions of slack water, rooted aquatic vegetation can grow. At time of flood, areas of terrestrial vegetation are also inundated. Correlated with this increase in physical diversity, there is an increase in the diversity of the riverine fishes, so that, in general, there is a positive correlation between the number of species and stream order within a drainage system (Lotrich, 1973).

As the gradient slackens, the flow rate in the main channel also slackens, although high rates can occur in some large rivers (Welcomme, 1985). Species that spend all or most of their life in the water column of the main channel have a streamlined herring-like body form adapted for near-continuous swimming (Figure 2.6f). In contrast, the areas of slack water, often associated with vegetation, are occupied by species whose body form is better adapted for manoeuvrability or rapid acceleration over short distances. Manoeuvrability is enhanced by a short, relatively deep body, and enlarged paired fins (Figure 2.6e). The ability to accelerate rapidly is enhanced by a relatively long body that has an approximately

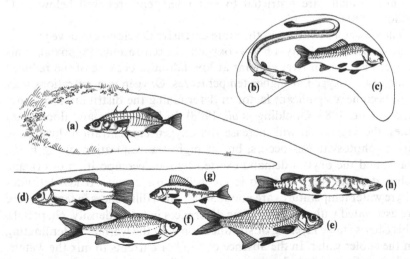

Figure 2.6 Representative fishes of potamon fish assemblage from western Europe: (a) threespine stickleback (*Gasterosteus aculeatus*) (55 mm); (b) eel (*Anguilla anguilla*) (500 mm); (c) carp (*Cyprinus carpio*) (400 mm); (d) tench (*Tinca tinca*) (300 mm); (e) bream (*Abramis brama*) (350 mm); (f) roach (*Rutilus rutilus*) (200 mm); (g) perch (*Perca fluviatilis*) (250 mm); (h) pike (*Esox lucius*) (500 mm).

equal depth along its length. This body form is epitomized by the pike, a piscivore, which lurks in vegetation until it can lunge at passing prey. In addition, there are species such as the catfishes, with dorso-laterally flattened bodies that are adapted for living on or close to the bottom, and fishes like the eel with long, slender bodies that can slither through vegetation.

Although the larger volumes of water tend to buffer temperature fluctuations to a greater extent than in the rhithron, in mid- and high-latitude rivers the annual fluctuations may still be large. The Mohave River in south-eastern USA has an annual fluctuation from 0.0–36.0°C (Castleberry and Cech, 1986), although this desert river is an extreme example. By contrast, the Rio Negro, a major tributary of the Amazon, has an annual temperature range of about 28–31°C (Goulding *et al.*, 1988). In shallow backwaters and flood plain lakes, the water temperature will tend to follow air temperature more closely than in the main channel and so be more variable. The species of mid- and high-latitudes that live in these flood-plain habitats are often more eurythermal than species living in the rhithron. Cyprinids such as the carp (*Cyprinus carpio*), the tench and the roach can survive in temperatures exceeding 30°C, whereas salmonids and other stenothermal species more characteristic of the rhithron usually are restricted to water temperatures well below 25°C (Elliott, 1981).

The slower flow rates and the more extensive development of vegetation increase the possibility of low oxygen concentrations (hypoxia). This danger increases in river systems at low latitudes because of the reduced solubility of oxygen at higher temperatures. Oxygen concentrations may then become a significant factor in determining the distribution of fishes (Welcomme, 1985; Goulding *et al.*, 1988). In backwaters and flood-plain lakes, the vegetation will increase the oxygen concentration during the day as photosynthesis occurs, but at night the respiration of the living plants and the oxygen demand exerted by the decomposition of organic debris depletes the oxygen to levels well below 2 ppm. At low latitudes where water temperatures are about 30°C, small differences in temperature are associated with relatively large differences in water density (chapter 1). This causes stable layering or stratification with the warmer water floating on the cooler water. In the absence of wind or currents to mix the waters, such stratification can develop even in shallow waters. The deeper water then has a tendency to become completely oxygen-free (anoxic). If subsequent rapid mixing of the water column takes place, many fish may be killed by the sudden deoxygenation of the surface waters, an effect exacerbated by the presence of toxic by-products of decomposition processes such as hydrogen sulphide (Welcomme, 1985).

At most latitudes, the fish of the potamon are more tolerant of low oxygen conditions than species of the rhithron. Indeed, some cyprinids can survive periods of deoxygenation by using anaerobic respiration and consequently can tolerate extremely low levels of oxygen (Hochachka, 1980). Other adaptations to low oxygen conditions have evolved, and are seen particularly but not exclusively in fishes of low latitudes.

The commonest adaptation is behavioural. As the oxygen concentration in the water drops, the fish start to come close to the surface and exploit the surface layer of water, which remains well-oxygenated by diffusion. This behaviour is called aquatic surface respiration (Kramer, 1983). A range of morphological traits such as upturned mouth, fleshy lips and a flattened head, enhance the ability to use this surface layer. In other species, adaptations allow the fish to obtain oxygen from atmospheric air (aerial respiration) (Johansen, 1970). The fish come to the surface and swallow air, which is then used for respiration. Structures modified to function as respiratory surfaces allowing aerial respiration include the skin (*e.g. Hypopomus brevirostris*), the lining of the mouth and pharynx (*e.g.* the electric eel, *Electrophorus electricus*), the swim-bladder (*e.g. Arapaima gigas*) and part of the alimentary canal (*e.g. Hoplosternum*). For most species, air breathing is supplementary to normal aquatic respiration but in a few species air-breathing is obligatory. The lungfishes of Africa and South America are obligate air-breathers and will drown if denied access to the water surface.

River water chemistry depends on the geology and soils of the catchment area but the distribution of fishes is only influenced by extremes in water quality. At one extreme are the blackwater streams and rivers of tropical rain forests. The water of the Rio Negro has been described as slightly contaminated distilled water (Welcomme, 1985; Lowe-McConnell, 1987; Goulding et al., 1988). These rivers are discoloured by organic compounds produced during the decomposition of plant materials. The pH is acidic, in the range 3.5–6.5. Yet such rivers can support a diverse fish fauna. Over 400 species have been collected from the Rio Negro and the full species count may approach 800. Only about ten Amazonian species are thought to be excluded from the Rio Negro, probably because of its hydrochemistry (Goulding et al., 1988). At the other extreme are streams in deserts that break up into pools during dry periods. Evaporation then increases the salt content of the waters. The resulting high salinities exclude primary freshwater species that are unable to osmoregulate in such conditions. A specialized and often highly localized fauna may then develop. The cyprinodont species of the genus *Cyprinodon*, living in saline conditions in arid, south western USA, have close relatives that live in estuarine and coastal environments. Some *Cyprindon* species can tolerate salinities in excess of 90 ppt (Naiman and Soltz, 1981).

The increase in species richness with the increase in stream order of rivers is usually by the addition of species to the assemblage. However, as the gradient decreases with often a correlated increase in average water temperature and decrease in average oxygen concentration, the dominant species in the fish assemblage change. For European rivers, faunal zones, which succeed each other in sequence from the head waters to the coastal plain, are described as trout zone, grayling zone, barbel zone and bream zone (Figure 2.7) (Huet, 1959). This represents the sequence from assemblages of fishes preferring the cool, well-oxygenated water of small, fast-flowing low-order streams to those that prefer the wide, sluggish, sediment-laden, high-order streams. The transitions between the zones are

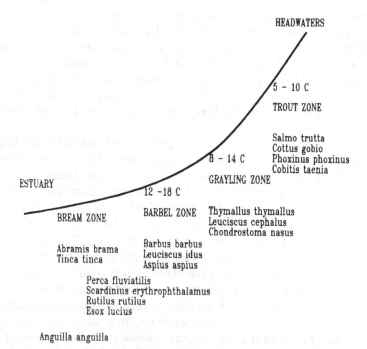

Figure 2.7 Huet's zonation of fish assemblages of European rivers. Representative species for each zone are given. C = temperature in °C.

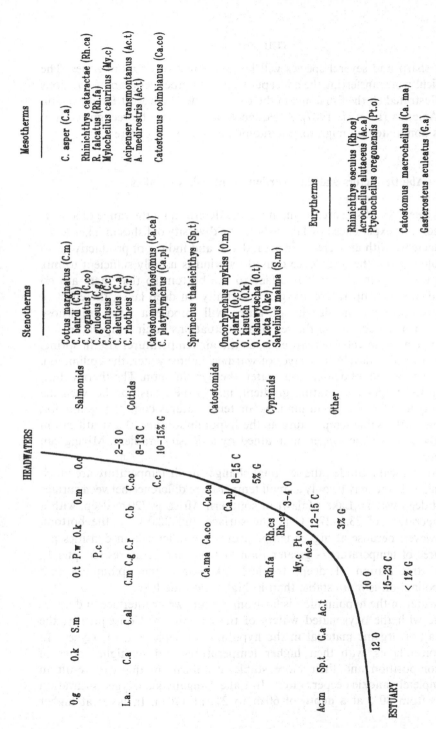

Figure 2.8 Zonation in rivers of Pacific north-west: 0 = stream order; C = temperature in °C; G = gradient. Species have been classified as steno- meso- or eurytherms. Based on Li et al. (1987).

not sharp and several species will be found in two or more zones. The difficulty of transferring the concept of zones to other geographical regions is illustrated by the fish fauna of the rivers of the north-west Pacific region of America (Li *et al.*, 1987). A sequence of zones can be recognized, but it is dominated by migratory salmonids and cottids (Figure 2.8).

2.4 Abiotic factors and the distribution of fishes in lakes

Lake depth and size are the main factors determining the range of abiotic conditions experienced by lake fishes. The diversity of fishes in lakes tends to increase with an increase in area, depth and indices of productivity. In shallow lakes, the turbulence caused by wind is usually sufficient to mix the water column from the surface to the bottom so that there are no gradients in temperature, oxygen or salinity. In deep lakes, the effect of temperature on the density of water will impose a structure whose permanence depends on the seasonal fluctuations in air temperature.

In temperate and high latitudes, as the air temperatures rise in spring, a thermocline develops. A layer of warmer, lighter water, the epilimnion, floats on a pool of denser, cool water, the hypolimnion. The thermocline, with its steeper temperature gradient, marks the transition between the two types of water. In autumn, as air temperatures cool, the epilimnion cools to the same temperature as the hypolimnion and the stratification of the waters is no longer maintained by a density gradient. Mixing can occur.

At tropical latitudes, the seasonal changes in air temperature are much smaller. There may be only a small temperature difference between surface and deep waters. Lake Malawi in southern Africa is 720 m deep, with a temperature of 23.5–27.5°C at the surface and 22.0°C at the bottom. However, because at higher temperatures the difference in densities per degree of temperature is greater than at lower temperatures (chapter 1), the stratification of deep, tropical lakes into the epilimnion and hypolimnion is more stable than in higher latitude lakes.

Water in the hypolimnion is far from the air–water interface and mixes little with the oxygenated waters of the epilimnion. Consequently, the decay of organic material in the hypolimnion depletes it of oxygen. In tropical lakes, with their higher temperatures and so higher rates of decomposition and their more stable stratification, this can result in completely anoxic deeper waters. In Lake Tanganyika, oxygen saturation falls from 80% at a depth of 60 m to 2% at 170 m. In lakes at higher latitudes, the hypolimnion is replenished with oxygen in the autumn when the breakdown of the thermocline allows mixing between the hypo- and epilimnion.

The increase in water density with an increase in salinity can, where there is a seepage of salt water into a lake, cause stratification with the lighter fresh water separated from the denser salt water by a halocline.

In relation to abiotic identities, a deep lake can be divided into four basic habitats. The shallow littoral at the lake edge will usually be well-oxygenated and show seasonal variation in temperature. The open water of the pelagic zone above the thermocline will be similar, but lack the structural variety provided by the substratum and macrovegetation of the littoral. Below the thermocline, will be cool pelagic waters and the benthic zone, which is sheltered from physical disturbances caused by the wind and from temperature fluctuations but prone to de-oxygenation.

Body shape in lake fishes is not determined by a need to cope with fast currents but still reflects the design constraints imposed by the hydrodynamic properties of water. In the shallow, littoral region, two body shapes predominate—both forms already met with in the slow-moving waters of flood-plain rivers. (Indeed, the two habitats often share many species.) Some species are deep-bodied, with well-developed pectoral and/or pelvic fins—a body shape for manoeuvrability. A good example is provided by the sunfishes, *Lepomis* spp., which are characteristic of the shallow littoral of lakes in north-eastern America (Figure 2.9). Others, the northern pike is an example, are elongated, with a body well-adapted for rapid acceleration. In the open pelagic waters, the species tend to be more streamlined with a narrower tail stalk and a notched tail, features characteristic of fishes that are good at cruising. The herring-like alewife (*Alosa pseudoharengus*) of the American Great Lakes is an example. In northern lakes, the benthic fishes include the sculpins (Cottidae) and the burbot with its broad, flat head and elogated body. These benthic fish pounce on their prey, relying on their ability to accelerate rapidly over short distances.

The relationship between body form and habitat is well illustrated by the dominant piscivorous fishes in Canadian lakes. In shallow lakes, the northern pike is dominant. In deeper, larger lakes the walleye (*Stizostedion vitreum*) also becomes important, while in the deepest and largest lakes, the lake trout (*Salvelinus namaycush*) is dominant (Marshall and Ryan, 1987). As Figure 2.10 shows, the body shapes in this series show a progression from the pike form in which the depth of body (including median fins) is similar along the length of the body with the

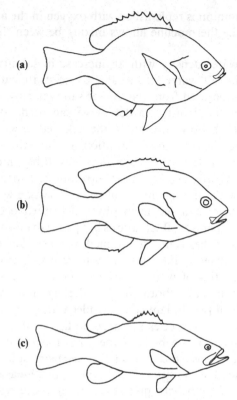

Figure 2.9 Centrachids of eastern North American lakes and streams: (a) bluegill sunfish (*Lepomis macrochirus*); (b) green sunfish (*L. cyanellus*); (c) largemouth bass (*Micropterus salmoides*).

median fins lying posteriorly, to the trout form in which the dorsal fin is more central, the tail stalk narrower and the tail fin more deeply notched.

Similar trends in body shapes are seen in tropical lakes. In the Great Lakes of Africa, the cichlids (Cichlidae) show a progression from deeper-bodied forms associated with the shallower, littoral areas to more streamlined forms living offshore.

In higher-latitude lakes, the presence of the thermocline in summer establishes marked differences in temperature between different areas of the lake. In Lake Michigan, the distribution of species caught in bottom trawls was studied in a region where the thermocline intersected the bottom and so generated steep gradients in temperature in the bottom waters (Brandt *et al.*, 1981). Different species achieved their maximum abundances

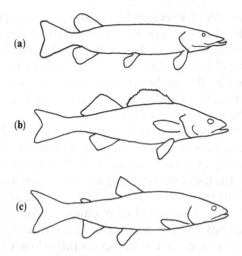

Figure 2.10 Dominant predators of Canadian lakes: (a) northern pike (*Esox lucius*), found in shallow lakes; (b) walleye (*Stizostedion vitreum*), found in deeper lakes; (c) lake trout (*Salvelinus namaycush*), found in the deepest lakes.

at different temperatures. During the day, young alewives were found in water warmer than 17°C, whereas the adults reached their maximum abundance at 11–14°C. Adult yellow perch and trout perch (*Percopsis omiscomaycus*) were most abundant at 15–16°C, while adult smelt (*Osmerus mordax*) were in water at 7–8°C. At night, the slimy sculpin (*Cottus cognatus*) was at its maximum abundance at 3–4°C. In temperate lakes, as in temperate rivers, two groups of fishes can often be distinguished. The warm-water species tolerate temperatures that reach the high 20s or low 30s, whereas the cold-water species are restricted to waters that do not exceed about 25°C. The latter may continue to feed actively when temperatures drop below about 10°C, whereas the former, even if they can tolerate low temperatures, cease to feed.

In tropical lakes, the temperature gradients are shallower except in the shallowest water or where hot springs are present. In Lake Magadi in East Africa, a cichlid, *Oreochromis grahami*, lives in water that reaches 39°C because of the presence of hot springs (Lowe-McConnell, 1987). Nevertheless, few species can tolerate such high temperatures.

In deep tropical lakes, much of the deeper hypolimnion lacks any permanent fish fauna because of the lack of oxygen. A few deep-water species may make excursions into these anoxic waters to feed before returning to oxygenated waters.

In lakes at higher latitudes, oxygen can become a limiting factor in two situations. In shallow lakes in summer, the high rate of decomposition of the vegetation may cause deoxygenation and the fish must then depend on the surface film (Gee *et al.*, 1978). In winter, a thick layer of ice and snow will cut the water off from the air. If decomposition occurs, the water deoxygenates and the fish cannot get to the surface film. Such conditions can result in the deaths of fish—called winter kills.

In deeper lakes, the volume of cold water will contain sufficient oxygen and because water is densest at 4°C, the deeper waters provide a refuge from the colder but ligher surface waters. At high northern latitudes, fish such as the Arctic Char (*Salvelinus alpinus*) will move from feeding grounds in coastal waters into fresh water to over-winter in temperature refuges. If they stayed in the sea, they would be in danger of freezing.

In most lakes, salinity will not be a factor in the distribution of fishes. However, in lakes that experience strong evaporative losses, salinities that exclude most freshwater species develop. If habitable at all, such lakes may only contain species that are derived from marine stocks and have retained the ability to osmoregulate in water that has a higher ion content than their body fluids. Lake Magadi, in addition to reaching over 40°C in places, also has a salinity of 40 ppt. The one species that lives in these conditions is a cichlid—a family with many relatives that are marine families.

In some lakes, the effect of salinity may be more subtle. Lake Tanganyika has experienced periods when water loss has been mainly through evaporation. This has produced water with an unusual ionic composition and concentration for a freshwater lake. Lowe-McConnell (1987) suggests that this has prevented some fishes colonizing the lake. Major families in Lake Tanganyika, the clupeoids and the centropomids, are predominantly marine families.

At the other end of the scale, an analysis of the fishes in Canadian lakes suggested that some species are unable to colonize lakes too low in calcium or too acidic (Henderson, 1985) (chapter 7). Nutrient-poor lakes in Wisconsin (USA) show a sequential loss of species correlated with an increase in the harshness of the abiotic conditions (Rahel, 1984). Lakes with a low pH (5.2–5.4) and which tend to have low oxygen concentrations in winter have, at most, only two fish species—the yellow perch and the mudminnow (*Umbra limi*). Those lakes that are less acidic but still suffer low oxygen in winter have an assemblage dominated by cyprinids. The lakes with more benign abiotic conditions have an assemblage characterized by centrarchids.

The waters of lakes as well as that of rivers is often discoloured by pigments of both aquatic and terrestrial origin and often carries a significant sediment load. Nevertheless, freshwater fishes active during the day (diurnal) or at dusk and dawn (crepuscular) and even some nocturnal species have good vision, which probably includes colour vision. Although many species are drab browns and olive greens, many also have red spots or blotches. In clear lake waters, fish colours include the blues and yellows seen in some cichlids. Thus there are correlations between the coloration of the fishes and the transmission qualities of the water (Lythgoe, 1979) (see page 10).

2.5 Abiotic factors and the distribution of fishes in estuaries

Estuaries represent a major challenge to any living organisms because of the wide range of abiotic conditions experienced in them. Foremost is the change in salinity down the estuary from freshwater conditions in the river proper to full sea water at the outer fringe of estuarine influence. Furthermore, salinity at any point in an estuary will change with the tides and with the pattern of discharge of fresh water from the river. The lower density of fresh water compared with sea water means that the river tends to flow over a tongue of more saline water stretching up the estuary, with the position of the tongue varying tidally and seasonally. Funnel-shaped estuaries increase the amplitude of the tidal variation and can also funnel winds, thus creating complex currents. This water movement, together with the sediment load brought down by the river, often makes estuaries turbid with a shifting, unstable substrate.

Tidal movements and seasonal floodings create periodically inundated pools and lagoons along the sides of the estuaries. These pools can experience large changes in salinity, oxygen concentration and temperature even over short periods.

In the tropics, mangrove swamps line many estuaries as well as sheltered coastlines with waters that have some estuarine characteristics, including variable salinity and a high sediment load (Lowe-McConnell, 1987). Within these swamps, abiotic conditions can be severe, with temperatures ranging from 20–40°C, salinities from 0 to 46 ppt and highly variable oxygen concentrations.

Perhaps not suprisingly, given the harshness of the abiotic conditions, the number of species that live out their whole lives within the estuarine environment is low (McClusky, 1989). However, estuaries and mangrove swamps are used by many coastal marine species as nursery areas, where juveniles grow towards maturity. Estuaries are also corridors for those species that pass regularly between fresh waters and the sea as part of their normal life-history (chapter 4).

Although some marine species may penetrate far up the estuary in the tongue of salt water, the only species found in full estuarine conditions are euryhaline species that are able to survive over the wide range of salinities experienced. Stenohaline marine and freshwater species are excluded. Typical estuarine species are gobies (Gobiidae), sticklebacks (Gasterosteidae), cyprinodonts (Cyprinodontidae) and atherinids (Atherinidae) (Figure 2.11). Many of these are small (> 100 mm) fish that spend all or part of their life in the fringing pools and lagoons. Groups that use estuaries as nursery areas include the herrings, anchovies and sprats (Clupeiformes) and mullet (Mugilidae). The juveniles of about 400 marine species have been collected from mangrove swamps, including young grunts (Haemulidae) and snappers (Lutjanidae).

Among the commonest estuarine fishes are benthic species. Gobies live in estuaries in both tropical and temperate zones. Many lack a swim-bladder and their pelvic fins are joined to form a 'sucker'. Some

Figure 2.11 Representatives of estuarine fish assemblage in western Europe: (a) common goby (*Pomatoschistus microps*) (60 mm); (b) flounder (*Platichthys flesus*) (250 mm); (c) grey mullet (*Mugil cephalus*) (250 mm); (d) juvenile herring (*Clupea harengus*) (100 mm); (e) juvenile whiting (*Merlangius merlangus*) (100 mm).

flatfish species (Pleuronectiformes) spend part of their life-cycle in estuaries. In European waters, the flounder (*Platichthyes flesus*) is sufficiently euryhaline to penetrate into fresh water. The flatfish also lack swim-bladders and rest on the bottom on either their left or right side. Their flattened body-shape allows them to accelerate off the substratum into the water column or to lunge at benthic prey. However, when swimming for extended periods, they take advantage of tidal currents by entering or leaving the water column at appropriate times (chapter 4).

Other common fishes of estuarine shallow waters, the sticklebacks and cyprinodonts, have swim-bladders and are more streamlined than the gobies. They also have well-developed pectoral fins, which they use for leisurely swimming and careful manoeuvering, only switching to using their body and tail for locomotion to dart over short distances.

In contrast, several of the species that use estuaries as nurseries, such as herring and sprat, have a streamlined body well adapted to cruising in open water.

The importance of salinity and temperature tolerance to the distribution of estuarine species is illustrated by a study of three species of snooks (Centropomidae), teleosts that live in coastal south-east Africa (Martin, 1988). The three species are distributed longtitudinally down estuaries. *Ambassis productus* lives in the upper reaches, whereas *A. gymnocephalus* lives at the mouths where the salinity is close to that of full sea water. *A. natalensis* occupies the intermediate and so most variable stretches of the estuaries with a distribution that overlaps those of the two other species. Experimental tests showed that these distributions correlated with the salinity and temperature tolerances of the three species.

The goby, *Gillichthys mirabilis*, illustrates the extreme euryhalinity of some estuarine species. In September, fish that were exposed to a salinity of 70 ppt (twice sea water) for 70 days still showed the expected seasonal increase in the size of the gonads (De Vlaming, 1971). The rate of recrudescence of the gonads was slower than in fish kept at 35 ppt.

Those species that occupy estuarine pools, lagoons and mangrove swamps can also experience low oxygen concentrations as organic material decays. As in fresh water, some species, like the temperate sticklebacks, resort to aquatic surface respiration. Others can use atmospheric oxygen. *G. mirabilis* starts to gulp air when the oxygen concentration in the water falls to about $2\,\text{mg}\,\text{l}^{-1}$. The respiratory surface is formed by vasculated areas in the mouth and pharynx (Todd and Ebeling, 1966).

Those species that can tolerate the estuarine conditions can reach high densities. In salt marsh pools in the St Lawrence Estuary, densities of sticklebacks average $20\,m^{-2}$ and can reach as high as 60 fish m^{-2} (FitzGerald and Whoriskey, 1985). Young gobies, *Pomatoschistus microps*, can reach densities of $75\,m^{-3}$ in the Tamar Estuary in south-west England (Dando, 1984). These densities are higher than would be expected in non-estuarine habitats.

2.6 Abiotic factors and the distribution of fishes in littoral and sub-littoral marine waters

In coastal waters affected by wave action, two basic types of habitat can be distinguished; the first is illustrated by rocky or coral reefs, the second by beaches and sub-littoral zones of mud, sand and shingle.

On reefs, the structural complexity provides a range of habitats—this is especially true of coral reefs. The coral-reef habitats include the adjacent off-reef floor, the reef slope or front, the reef surface, the reef flat and the lagoon protected from wave-action by the reef (Goldman and Talbot, 1976; Lowe-McConnell, 1987). The lagoon often contains stands of sea grass, which act as nursery grounds sheltering juveniles and as feeding grounds for some herbivorous reef fishes (chapter 5). This architecture is created by the reef-building coral, which require warm ($> 20°C$), fully saline water that is clear and free from sediment. The coral have a symbiotic relationship with photosynthetically active algae living in the corals' tissue, hence the requirement for clear water. Reef building is most active in water less than $30\,m$ deep. The requirements of the coral mean that, in terms of water temperature, salinity, sediment load and other abiotic identities, coral reefs are benign in comparison with the rocky reefs in more temperate climates. However, both rocky and coral reefs will be subjected to variation in abiotic identities in zones that are uncovered when the tide recedes and in any shallow, lagoon waters sheltered by the reef. These habitats are subject to water loss by evaporation, or dilution by rain water or run-off from the land.

The physical complexity of a reef provides shelters within which fish can, at least partly, avoid the effects of the turbulent water. On a sandy shore, the substrate itself is unstable, being swept about by the turbulence. Any pools that are left behind as the tide recedes are shallow and impermanent.

Rocky and coral reefs support diverse fish faunas, many members of which live out most or all their life there. Indeed, the diversity of species on coral reefs in south-east Asia is among the highest observed for any aquatic habitat, when similar areas are compared (Sale, 1980). Well over 2000 species are recorded from the Philippines. This diversity is clearly related to the range of habitats provided. In one census at a site on the Great Barrier Reef in Australia, only 7% of the species were found in all reef habitats, while 49% were restricted to one or another habitat. Many coral reef species show strong habitat selection and recognize only a part of the reef as their normal living area (Goldman and Talbot, 1976). In contrast, the sandy littoral and shallow sub-littoral zones support a less-diverse fauna and one in which most members move into and out of in a way that avoids the worst effects of turbulence, although a few species may survive by burying themselves in the bottom (Gibson, 1986).

On reefs, the effects of turbulence are countered by two types of morphological adaptation (Figure 2.12). Most of the fishes that live there have small body sizes, species greater than 200–300 mm in length are rare. This allows them to seek shelter in crevices, holes, hollows and between the fronds of sea weeds. Secondly, body form reflects the tendency for these fishes to live on the bottom, avoiding, except at high tide, the turbulent water column. Families like the Cottidae, found in the fast-moving waters of headwater streams, also have representatives in the

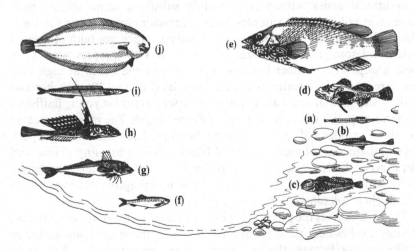

Figure 2.12 Representatives of fish assemblages of rocky (a–e) and sandy (f–j) beaches of western Europe: (a) broadnosed pipefish (*Syngnathus typhle*) (180 mm); (b) fifteenspined stickleback (*Spinachia spinachia*) (120 mm); (c) shanny (*Blennius pholis*) (150 mm); (d) father lasher (*Myoxocephalus scorpius*) (180 mm); (e) ballan wrasse (*Labrus bergylta*) (300 mm); (f) sprat (*Sprattus sprattus*) (100 mm); (g) lesser weaver (*Trachinus vipera*) (120 mm); (h) dragonet (*Callionymus lyra*) (200 mm); (i) sand eel (*Ammodytes tobianus*) (150 mm); (j) sole (*Solea solea*) (150 mm).

reef littoral at high latitudes. In some bottom-dwellers, like the blennies (Blennidae), the body is elongated and somewhat compressed. In others, the body is depressed as in the sculpins (Cottidae) and the clingfishes (Gobiesocidae). These bottom-dwellers are negatively buoyant because they have reduced swim-bladders or lack them altogether. Various modifications increase the friction between the fish and the substrate. In the gobies, the clingfishes and the snailfishes (Liparidae), the pelvic fins are modified to act as attachment organs. In blennies, the rays of the anal, pectoral and pelvic fins are hook-shaped, helping the fish to cling to the substrate.

In deeper water, where the turbulence is less, live deep-bodied species adapted to manoeuvering their way through the maze of the reef. On coral reefs, forms include the butterflyfishes (Chaetodontidae), the damselfishes (Pomacentridae), wrasses (Labridae) and the surgeonfishes (Acanthuridae). Deeper-bodied forms also move into the littoral reef zone at high water when turbulence is least, retreating to the sub-littoral as the tide recedes. Pelagic fishes with streamlined bodies cruise in waters immediately adjacent to the reef including both zooplanktivores such as clupeoids and piscivores like jacks.

In littoral zones with a soft, mobile substrate some species bury themselves at low tide. Examples include representatives of the sand eels (Ammodytidae) and the weavers (Trachinidae), which are both elongated forms, and some flatfishes (Pleuronectiformes). Other representatives of these groups and of other families move in over the beach at high tide, but retreat to deeper water as the tide recedes (Figure 2.12). Beaches and shallow sub-littoral areas are important nursery areas for young flatfishes, even those species whose adults live in deeper water. The major exception to these modes of life on soft beaches are the mudskippers, Periophthalmidae, of muddy tropical beaches. These are amphibious and can survive the low tide exposed to the air so long as they can remain moist. Mudskippers are also common in many mangrove swamps, which reflects their tolerance of extreme physical conditions.

Temperature has the potential of being a major factor in the distribution of littoral fishes because at the top of the shore, the water temperature in shallow pools follows the air temperature more than in the deeper sub-littoral waters (Gibson, 1982). On a Hawaiian shore, temperatures in excess of 40°C were recorded in pools occupied by blennies and gobies. At higher latitudes, upper shore pools can have water temperatures below zero in winter. Even within a day, the temperatures in shore pools can change rapidly when the tide advances as cooler water floods into pools warmed by the sun. The lower limit of reef fishes may also be determined by temperature. Johannes (1981) describes the behaviour of fish being chased down the slope of a coral reef in the western Pacific. At about 30 m, the fish suddenly stopped and went no deeper. The fish had reached the thermocline and would not swim into the cooler waters.

Salinity may also change rapidy in littoral pools (Gibson, 1969, 1982). Along the coast of Texas, salinities in excess of 80 ppt can develop. Such pools are occupied by extremely euryhaline species like the goby, *G. mirabilis* and cyprindonts like *Cyprinodon*. More typically, in littoral species like the blenny *Blennius pavo*, the tolerance of high salinities is lower but still well in excess of the salinity of sea water. Littoral species are also often tolerant of low salinities, surviving indefinitely in salinities one quarter that of sea water. On coral reef systems, the species that live at the landward edge tend to be the euryhaline gobies, blennies and mullets (Lowe-McConnell, 1987).

Distribution on the shore will also be related to resistance to desiccation when the fish is emersed in air. In high humidities, some littoral fishes can survive for many hours out of water, although losing body water. The temperate-shore blenny, *Blennius pholis*, can survive 4 to 6 days of emersion while suffering a water loss of 22% of its body weight.

In fresh waters, estuaries, mangrove swamps and stands of sea grass, low oxygen conditions often develop because of organic decomposition. Water turbulence in the littoral zone usually prevents such conditions developing although the respiration of tide pool algae can reduce oxygen to low levels during the night. However, many littoral species can use aerial respiration, enabling them to survive lengthy periods exposed to the air when the tide recedes (Gibson, 1982). The skin, gills and the epithelium of the mouth and pharynx are the commonest respiratory surfaces and, as in the mudskippers, all three may be used.

A feature of coral reefs is the good light penetration—particularly for the blue end of the spectrum. Correlated with this is the presence of many brightly coloured teleosts, with blues and yellows often their prominent colours. The functions of such coloration is still a subject of debate and will be considered further in chapter 3.

The normal course of events on reefs can be interrupted by catastrophic changes caused by adverse weather conditions. Some coral reefs are in climatic zones that are prone to hurricanes. These can cause major disruptions to the physical structure of the reef. Hurricane Allen struck Jamaica in August 1980. It exposed large amounts of substratum on the reefs of the north coast through abrasion, fracture and the death of corals

(Woodley *et al.*, 1981). The threespot damselfish (*Eupomacentrus planifrons*) showed changes in distribution because the algal lawns they normally defended (chapter 4) were destroyed. The fish resumed such a defence 2–9 days after the hurricane.

In terms of their adaptations to abiotic factors, many littoral fishes, like those of the estuaries, are tolerant of wide-ranging conditions. This allows them to exploit a habitat from which other less-tolerant species are excluded or can only visit on a temporary basis. However, those reef fishes which by their movements avoid being exposed to unfavourable abiotic conditions at low tide, are often intolerant of such conditions.

2.7 Effect of abiotic factors on the distribution of open-sea species

From the coast, the continental shelf extends outward, with water depths increasing to about 200 m. Beyond this depth, the continental slope descends to the abyssal depths of 4000 m and more. The mean depth of the Earth's seas is about 4000 m, which shows that most of their volume lies above these great depths.

Away from the shore and shallow sub-littoral waters, abiotic identities such as temperature, oxygen and salinity show less variation, although sufficient to impose some dynamic structure on the environment. Other abiotic factors can also be important. For demersal fishes living on (benthic) or near (benthopelagic) the bottom, the nature of the substratum is important. In the deep oceans, pressure will restrict the distribution of some species. For both pelagic and bottom-living fishes, light penetration is also a factor.

In the sea, a shallow, warmer, wind-mixed surface layer lies like a skin on a vast volume of cold, deep water (Marshall, 1979). The zone between the two layers is marked by the thermocline, in which the temperature gradient is higher than above or below. At mid and low latitudes, the temperature of the surface layer in summer is between 15 and 30°C. Between depths of about 300 to 1000 m, the temperature drops to about 5°C but then with further increase in depth falls gently to 1–2°C at 4000 m. In the Arctic and Antarctic, there is little change in water temperature from surface to bottom, and the shallower water can be colder than the deeper waters dropping to temperatures below 0°C. In climatic zones in which there are large seasonal changes in air temperature, a secondary seasonal thermocline develops in summer separating the shallowest surface waters that are more strongly warmed from deeper waters above the permanent thermocline. This thermocline breaks down as air temperatures decline in autumn. In sub-tropical and tropical seas, a shallower thermocline separating the warmest surface waters with temperatures in excess of 20°C from the cooler underlying waters is virtually permanent. This shallow thermocline tends to be deeper in the west of oceans than in the east (Longhurst and Pauly, 1987).

Throughout most of the oceans, salinity gradients are extremely shallow. Exceptions occur where rivers bring large volumes of fresh water down to the sea, resulting in an 'estuarization' of the coastal waters, or where less saline water flows out from one sea into another. Relatively fresh surface water flows out of the Baltic Sea and northwards along the coast of Norway.

Surface waters, mixed by winds, are well oxygenated. Deep waters depend for their oxygen on convergence zones where surface waters rich in oxygen sink downwards. Because of the poor light penetration, the oxygen in the deeper waters is not augmented by photosynthesis, but is used up by the respiratory activities of living organisms and the decay of dead ones. Oxygen minimum layers develop in oceans. Between 10° and 20°N in the eastern Pacific, concentrations as low as $0.1 \, ml \, l^{-1}$ occur at depths between 200 and 1000 m (Marshall, 1979). An oxygen minimum layer is strongly developed in the northwest Indian Ocean. At higher latitudes and in deeper waters, oxygen concentrations of $5-6 \, ml \, l^{-1}$ are more typical (cf. Figure 1.3).

Over the continental shelf, there are high levels of suspended and dissolved material in the water. Light penetration is poor with the longer wavelengths, the reds and greens penetrating further than the short (blue) wavelengths. Under these conditions, visible light penetrates to only about 200 m. In the open ocean, where levels of both inorganic and organic particles are lower, blue light may be detectable at depths approaching 1000 m (Smith, 1976; Lythgoe, 1979).

In the open sea, water movement is not a major physical factor in determining optimal body shape. Rather, body shape will reflect the essentially unbounded nature of the environment except at the sea bottom.

Fish that live in the pelagic zone from the thermocline to the surface have, for the most part, body shapes adapted for continuous cruising (Webb, 1984) (Figure 2.13). The characteristic species include the plankton-feeding clupeoids, the carnivorous mackerels and tunas (Scombridae), jacks (Carangidae) and marlins (Istiophoridae) with, at the top of the food web, large predatory pelagic sharks (chapter 5). All these species are streamlined with a narrow tail stalk and a forked or

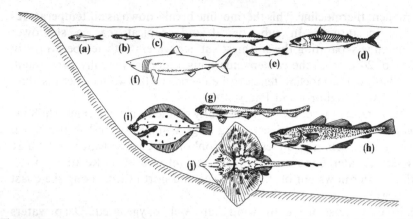

Figure 2.13 Representatives of the pelagic (a–f), demersal (g–h) and benthic (i–j) fish assemblages of the continental shelf of western Europe: (a) lesser argentine (*Argentina sphyraena*) (180 mm); (b) anchovy (*Engraulis encrasicolus*) (150 mm); (c) gar (*Belone belone*) (700 mm); (d) mackerel (*Scomber scombrus*) (300 mm); (e) herring (*Clupea harengus*) (240 mm); (f) basking shark (*Cetorhinus maximus*) (8 m); (g) dogfish (*Scyliorhinus caniculus*) (700 mm); (h) cod (*Gadus morhua*) (650 mm); (i) plaice (*Pleuronectes platessa*) (400 mm); (j) skate (*Raja batis*) (1 m).

crescent-shaped tail fin and relatively small or inflexible paired fins. Some of the larger of these active pelagic teleosts lack a swim-bladder so like the sharks they would sink if they stopped swimming. A curiosity of the pelagic zone are the flying-fish (Exocoetidae). Although streamlined, they have enlarged pectoral fins and an enlarged lower lobe to the tail fin. They are able to escape predators by leaping out of the water and gliding for many metres, using the paired fins as wings. Another curiosity, and an exception to the usual body shape is the sunfish (Molidae). This large (up to 3 m in length) pelagic teleost has a short, deep body and no tail stalk with the tail fin reduced to a leathery fold. The large triangular dorsal and anal fins are set well back along the body. Sunfish feed on jellyfish, comb jellies and other invertebrates whose capture does not require the ability to chase the prey down.

Demersal fishes usually have body shapes better adapted for swift darts to catch prey than for continuous cruising (Figure 2.13). Two common body forms among benthic fishes are a flattened body exemplified by rays and flatfishes or an elongated body exemplified by eelpouts (Zoarcidae). Benthic teleosts that spend part of their life resting on the bottom or in burrows usually lack a swim-bladder and so are heavier than water. In this they resemble the rays and skates with which they share the benthic habitat. Benthopelagic fishes, such as the cod, are more streamlined; however the number and position of the median fins maintain the depth of the body along its length and, with a tail fin that is only slightly, if at all, notched, are suggestive of fish that can accelerate quickly. At great depths, the commonest demersal fishes are the rattails (Macrouridae), with large heads and elongated tails. On the bottom, in the deepest waters of the sub-tropical and tropical oceans, are forms like the tripod fish (Chlorophalmidae), which support themselves on the substrate with strengthened fin rays (Marshall, 1979). Many benthopelagic teleosts have swim-bladders providing neutral buoyancy. Benthopelagic sharks have enlarged livers containing low-density oils, which means the sharks approach neutral buoyancy. This will allow the fish to hover or cruise slowly above the substrate, before darting down to take food off the bottom.

Away from the continental shelf, waters between 200 m and 1000 m deep form the mesopelagic zone. The fish fauna is dominated by two groups of teleosts, the lantern-fishes (Myctophidae) and the stomiiformes (Marshall, 1979). The former are usually small, spindle-shaped fishes with deeply forked tail fins, although with relatively large heads. Many species of lantern-fish show extensive daily vertical migrations (chapter 4). Most lantern-fishes have a functional swim-bladder. Somewhat similar in shape but more elongated and with a smaller mouth are the argentines (Argentinidae). The stomiatoid fishes of the mesopelagic zone show a range of body forms. The hatchet-fish (Sternoptychidae) are typically small, with deep, laterally compressed bodies. These also have a functional swim-bladder. Many of the larger, predatory fishes of the mesopelagic zone, including the larger stomiatoid fishes have an elongated body shape. In the dragonfishes (Melanostomiidae), the dorsal and anal fins are set well back along the body. In other forms such as the lancet-fish (Alepisauridae), the dorsal fin is set more anteriorly. Most of these larger mesopelagic forms lack a swim-bladder.

In the lower mesopelagic zone and below that (deeper than 1000 m) in the bathypelagic zone, the most diverse group is the ceratioid angler-fishes (Lophiiformers). These, and other bizarre-shaped fishes such as the gulper eel illustrate the extent to which body form has become emancipated from the constraints imposed by the need to maintain a position in a current or cruise at moderate speeds over long distances. In these deep-sea forms, living as they do in an extremely unproductive environment, the food-gathering apparatus of a huge mouth and distensible stomach has heavily compromised the expression of the locomotory apparatus.

Thermal tolerance in marine fishes ranges from forms that are extremely stenothermal to those that are eurythermal. In parts of the Antarctic Ocean, fishes may live their whole lives at temperatures below 0°C (Everson, 1984; MacDonald et al., 1987). The fauna is dominated by the nototheniids, large-headed, perciform teleosts. Nototheniids taken from McMurdo Sound survived indefinitely at 4°C, but higher temperatures were lethal. Naturally, the fish may rarely experience temperatures higher than − 1.4°C. These Antarctic fishes and species that live in the Arctic seas have anti-freeze molecules in the blood, which reduce the freezing point of body fluids (DeVries, 1971). In the absence of the anti-freeze molecule, the freezing point of the blood of the nototheniids would be almost 1°C higher than that of the sea. In nototheniids, glycoprotein molecules form the antifreeze while in Antarctic eelpouts, it is formed by peptides.

In contrast, those species that migrate vertically from below or in the thermocline to the surface waters (chapter 4) must be able to tolerate marked temperature changes even over a 24 h period. Off the coast of Norway, Atlantic herring, Clupea harengus, may descend to 300–400 m during the day moving into warmer surface waters at night. In these cold, northern waters the fish are moving from temperatures of 3–4°C into water at 5–7°C every 24 h (Harden Jones, 1968). In tropical and sub-tropical waters, many species of lantern fishes migrate from cold mesopelagic waters into warm epipelagic waters at night where they actively feed. Conversely, some of the large epipelagic predatory fish such as the marlins and swordfish, will dive deep into the cold sub-thermocline waters, traversing a temperature range from above 25°C to 8°C, apparently on feeding forays.

Epipelagic fishes in temperate and sub-polar seas must also be sufficiently eurythermal to tolerate the seasonal changes in water temperature, although in the summer months they can select cooler water by moving into or below the seasonal thermocline.

On a wide geographical scale, the distribution of many fishes can be correlated with water temperature. The distribution of Sparidae along the west coast of Africa provides a good example (Longhurst and Pauly, 1987). Several species extend from the cooler northern waters (Mauretania) to the cooler southern waters (Angola). In the tropical waters off the Gulf of Guinea, the sparids are found in cooler, sub-thermocline waters. In this region, the thermocline is shallow with the warmest, surface mixed layer only 20–40 m deep, and light can penetrate well into the sub-thermocline waters. In contrast, on the other side of the Atlantic, along the eastern coast of the Americas, the thermocline is deeper. There are northern and southern communities of sparids but, in the central tropical regions, the habitat occupied by sparids in the eastern Atlantic is occupied by the tropical pomadasyids.

Most marine species will experience only slight changes in salinity and are probably stenohaline. However, species that live in coastal areas, especially areas influenced by freshwater run-off are euryhaline. This is illustrated by three clupeid species off west African (Blaxter and Hunter, 1982). *Sardinella aurita* is stenohaline and is not found in salinities below 35 ppt. *S. cameronensis* can survive in salinities down to 20 ppt, while *S. eba* is able to enter estuaries and lagoons.

In some clupeoid species, the larvae and young stages that use estuaries as their nursery areas are more tolerant of low salinities than older stages.

Many of the species found in the mixed epipelagic zone are highly active fish with a correspondingly high metabolic rate. Species such as the tuna and dolphin (Coryphaenidae) have a high gill surface area (Hughes, 1984) and are intolerant of low oxygen concentrations. Under normal circumstances they would never experience such conditions.

Although some fishes are caught in the oxygen-minimum layer found in sub-tropical oceans, in the eastern Pacific between 10°C and 20°C, several species of lantern fishes are absent, probably because of the poor supply of oxygen. In the Arabian Sea, there is a layer in which oxygen concentrations as low as $0.08\,\mathrm{ml\,l}^{-1}$ may be reached. The snaggletooth, *Astronesthes lamellosus* (Astronesthidae), lives in this oxygen-poor environment but differs from related species in having gill filaments three times longer (Marshall, 1979).

In contrast, in cold Antarctic waters the oxygen concentration is sufficiently high to have allowed the evolution of fishes that lack haemoglobin. The Antarctic Channichthyidae rely on oxygen carried in simple solution in the blood serum. Antarctic fishes with haemoglobin have an oxygen-carrying capacity ten times that of the Channichthyidae, but the latter have a large blood volume, a large heart and show high rates of gill ventilation (MacDonald *et al.*, 1987).

For demersal fishes, habitat selection based on the nature of the substrate may play an important role in determining the species present and their relative abundances in an area. On the continental shelf of West Africa in the 1960s, six assemblages could be defined by the water depth and the physical characteristics of the bottom (Longhurst and Pauly, 1987). Distinctive assemblages of fishes living on or near the bottom have also been identified on other continental shelves including the eastern coast of South America off Guiana (Lowe-McConnell, 1987), and off the eastern seaboard of North America (Horne and Campana, 1989).

Fishes living in deep waters experience high pressures (approximately

1 atmosphere for every 10 m depth). Only those species with the appropriate adaptations to their physiological and biochemical systems can
survive and reproduce in these high pressure regimes (Hochachka and
Somero, 1984).

The vertical distribution of oceanic fishes correlates with the penetration
of light. Epipelagic fishes have well-developed eyes. However, mesopelagic
fishes also have eyes that are normal to very large although they are living
in waters lit only by a tiny proportion of the light reaching the ocean
surface (Marshall, 1979; Lythgoe, 1979). The visual pigments in these fishes
tend to absorb blue–green wavelengths (480–485 nm) most effectively. This
correlates with the wavelength of the dim downwelling light that penetrates
most deeply in clear, oceanic waters. Many mesopelagic fishes, such as
the lantern fishes, are also bioluminescent, emitting light that is characteristically blue–green. This bioluminesence may function as a form of
camouflage (chapter 3), but may also be important in attracting and
recognizing mates (chapter 6).

In bathypelagic species and benthic species living at depths greater than
1000 m, eyes tends to be reduced in size. In those benthic fishes living
higher up the continental slope, the eyes are often well-developed and
show many of the adaptations found in the eyes of mesopelagic fishes
(Marshall, 1979).

CHAPTER THREE

BIOTIC FACTORS AND THE STRUCTURE OF FISH COMMUNITIES

3.1 Introduction

The composition of a fish assemblage at a given locality is the result of the operation of a series of filters (Tonn *et al.*, 1990). The process of speciation determines the total complement of species that could be present. Biogeographical processes determine those species that have access to the locality. Abiotic conditions define which species could establish self-perpetuating populations. The productivity of the locality must be sufficient for populations to reach a density at which they are self-sustaining. Finally, on a local scale, biotic interactions and habitat selection may determine the details of the distributions of species across a range of habitats and microhabitats that are abiotically suitable (Figure 3.1).

The biotic interactions experienced by an individual may be with members of the its own species (conspecific interactions), or with other fish species or with other organisms, which may range from microorganisms to large, warm-blooded vertebrates (heterospecific interactions). In terms of its survival or its reproductive success, the fish may benefit or lose because of these interactions. The nature of the interactions will depend on many factors (Begon *et al.*, 1989). These include the diversity of other organisms with which the fish shares the habitat, the physical structure and abiotic conditions that characterize the habitat, the size of fish and the stage it has reached in its life cycle.

3.2 Classification of the interactions

3.2.1 *Predation*

A fish both eats and is in danger of being eaten. The food eaten is either living organisms or the detritus produced by living organisms. This aspect

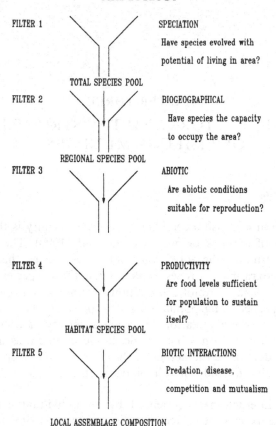

Figure 3.1 Simple conceptual model of factors determining the species richness at a locality. Loosely based on Tonn *et al.* (1990).

is discussed in chapter 5. Many predators, ranging from coelenterates to cetaceans, prey on fishes. The size of the fish is a major factor determining what can eat it and so the risk of death from predation changes as the fish grows. This is particularly true of teleost fishes, which produce small eggs (chapter 6) and consequently small and vulnerable larvae and young juveniles. The risk of predation will also depend on the density of predators, their rate of food consumption and on the possibility of evading their attacks (Taylor, 1984). Cannibalism is common among fishes. In some cases, this involves adults or large juveniles eating eggs, larvae or small juveniles (inter-cohort cannibalism), but it can also involve fish eating conspecifics of approximately the same age (intra-cohort cannibalism).

3.2.2 *Pathogens*

Fish may also be killed or debilitated by infectious organisms (Roberts, 1989). These will include micropathogens such as viruses, bacteria and fungi or macropathogens such as flukes and tape worms (Platyhelminthes), round worms (Nematoda), spiny-headed worms (Acanthocephala) and parasitic Crustacea like the sea louse. The effects of pathogens are especially obvious in fish farms where large numbers of fish are kept together so the risk of infection is great. Heavy mortalities caused by pathogens occur in natural fish populations, although the usual rate of mortality from this cause is not well-known. Debilitation caused by pathogens may increase the risk of predation or reduce the capacity to compete effectively. Often, the effects of pathogens only become obvious when they cause visible lesions on the fish such as those that characterize the disease of ulcerative dermal necrosis (UDN) in salmonid fishes.

3.2.3 *Competition*

The survival or reproductive success of a fish may also be decreased by the activities of other organisms that reduce the access of the fish to a resource that it requires. In intraspecific competition, the other organisms are conspecifics whereas, in interspecific competition, the other organisms are heterospecifics. The competition may occur because the other organisms are depleting the resource that is required. In this case the competitors may never meet, but compete because they exploit the same resource (exploitation competition). Competition may also occur because the competitors interfere directly with the activities of the fish, for example by excluding the fish from a feeding area or shelter by aggressive behaviour (interference competition) (chapter 4).

   Where two or more species that exploit a smilar range of resources show some degree of habitat segregation when in close proximity, this segregation may be caused by present day competitive interactions (interactive segregation). Conversely, it may be a result of the differing habitat preferences of the species (selective segregation) rather than competitive interactions.

3.2.4 *Mutualism*

Competitors, predators and pathogens increase the risk of death and decrease the possibility of successful reproduction but some interactions may have the reverse effect, increasing the fitness of the actors. Fishes take part in several such mutualistic interactions including cleaning interactions involving either a fish or a shrimp as the cleaner removing external parasites from the client fish.

3.3　Role of biotic factors and community structure

The study of community structure has two main themes. The first is the analysis of the factors that determine the number of different sorts of organisms that are found together in a habitat and their relative abundances. The second is the analysis of the rate at which energy and nutrients are used within the habitat and the biomass and size-structure of the organisms that are present. The two themes are related but the nature of the relationships is still obscure. Often interest centres on a taxonomic group and so the questions can be asked: what factors determine the number and abundances of fish species co-existing within a habitat, and what determines the biomass and size-structure of the species present? In other words, the focus is on the fish assemblage rather than the whole community.

Answers to these questions will require information about the effects of abiotic and biotic conditions on the species present and on the relative importance of abiotic conditions *versus* biotic interactions.

If interactions with natural enemies (predators, pathogens and competitors) restrict the species to a sub-set of the ranges of the abiotic identities that define its fundamental niche (chapter 2), then its spatial distribution will be less than that predicted from the fundamental niche. This restricted distribution is defined by the realized niche of the species. Much of the analysis of fish assemblages is concerned either directly or indirectly with the problem of the extent to which the observed distributions and abundances of species reflect their fundamental or their realized niches. To what extent are observed distributions determined by biotic interactions rather than the abiotic conditions?

3.4　Biotic interactions and community structure in rivers

The number of fish species present in a river drainage system increases with stream order. Abiotic identities such as current velocity tend to vary more in lower-order than in higher-order streams. These two observations suggest that the importance of biotic interactions in determining the distribution and abundance of fishes could increase downstream (Zalewski *et al.*, 1990). Even within the rhithron, the effect of abiotic and biotic factors will vary because of the riffle and pool morphology (chapter 2). In Jordon Creek, a second-order, warm-water stream in Illinois, the downstream region has well developed riffles and pools, whereas in an unmodified upstream stretch the riffles and pools are only moderately developed (Schlosser, 1982, 1987). Physical conditions vary less in the downstream than in the upstream stretch. In the upstream regions, the fish fauna is dominated by small cyprinids, whereas sunfishes (Centrarchidae) and suckers (Catostomidae) dominate downstream. Some of the species in the deeper, larger pools are predators of small fish.

On the basis of his study of Jordan Creek and similar studies in North America, Schlosser (1987) has suggested a preliminary model of the factors affecting the structure of fish communities in small streams. The crucial factor is the development of large, deep pools that can be occupied by large-bodied species including piscivores. Deep pools also act as refuges in periods when abiotic conditions are unfavourable as in winter or at times of high rates of discharge.

Stream stretches lacking such pools are low both in the number of species and in fish density. They are occupied principally by small-bodied forms—particularly juvenile cyprinids (or in other geographical regions, their equivalent). Large, piscivorous fish avoid such stretches. Small fish can find refuges from predatory birds by diving down between or under stones. These small fish often have rather transparent bodies or are cryptically coloured. Consequently, predation may be less important as a biotic interaction than further downstream. The small fishes may show some intra- and interspecific competition because they are all of a similar size and are exploiting similar foods. However, these assemblages of small fishes are susceptible to the effects of changes in the abiotic conditions. The maintenance of these shallow water communities will largely depend on repeated colonization from downstream. Many riverine species do show an upstream migration to spawn (chapter 4), which results in the recruitment of young fish to the upstream, shallow water community.

With the development of deeper pools, the number of species present increases, but there is also a shift towards fewer, larger individuals. Predation becomes an important interaction as smaller fish are restricted to shallow waters by piscivores such as the bass (Centrarchidae) found in deeper pools in many North American streams. This effect of predation on the distribution of smaller fish may increase the intensity of competition between them by confining them to limited regions of the stream.

Habitat segregation is a mechanism that will reduce the intensity of competition. Gorman (1987) found that an assemblage of small cyprinid species in a fourth-order stream in Missouri showed a striking pattern of vertical and horizontal distributions at the edge of the stream. The six species were largely occupying different zones in the stream. The young of the year occupied different positions in the stream from the adults and so were effectively segregated from the latter. These cyprinids were mostly absent from large areas of the stream, which consisted of pools and which were occupied by large-bodied, piscivorous centrarchids. In streams in Panama, herbivorous catfish are segregated by size—the smaller fish live in the shallower parts of the streams whereas the larger catfish occur in the deeper pools. Predation by birds and other terrestrial piscivores is probably the major cause of this distribution. Small catfish can escape from such predation by hiding between stones and in debris in the shallow sections. The larger catfish can avoid predation by staying in deeper water (Power, 1987).

The distribution and abundance of fishes in the rhithron is prone to disruption by adverse abiotic conditions, including unusually heavy flooding or periods of drought. Several studies on streams in North America suggest that, after such disturbances, the species composition and even the abundances of individual species can recover to levels similar to the pre-disturbance levels (Ross *et al.*, 1985; Meffe and Sheldon, 1990). This recovery depends on the colonization of the affected areas by fish from less disturbed parts of the stream, and over a longer period, by natural recruitment through reproduction. The strong tendency of species to select particular habitats within the stream defined by depth and current velocity probably accounts for the reconstitution of the pre-disturbance community structure (Meffe and Sheldon, 1990). Good evidence for the role of biotic interactions in controlling the redevelopment of community structure is lacking.

Studies on the salmonid communities of more northerly streams also suggest that differences in habitat selection, interspecific interference competition for suitable resting and feeding stations and the need to avoid exposure to terrestrial predators all play a role in determining the relative abundances of species (Hearn, 1987).

Only species that through evolutionary time have evolved traits that allow them to recolonize disturbed sections of the stream are likely to persist in the rhithron. The persistence of the species composition and their relative abundances in a stream may depend on the regularity and timing of the disturbances. The structure of the fish communities in streams in which disturbances occur unpredictably both in timing and in size may change more over time than in streams in which the disturbances occur with more predictability (Grossman *et al.*, 1982; Moyle and Vondracek, 1985).

The higher number of species in the downstream fish assemblages increases the potential for a variety of biotic interactions, but the difficulties of sampling large rivers quantitatively has meant that much less is known about the structure of such assemblages than in low-order streams.

Studies on the diet of flood plain fishes in the Amazon basin suggest that most species take a wide range of foods and there is considerable overlap in the diet of many species (Goulding, 1980; Goulding *et al.*, 1988). This raises the possibility of interspecific competition for food when it is in short supply. A relatively high proportion of the species (35%) in the Rio Negro included fish in their diets, although only about 20 species out of 250 sampled were mainly fish-eating. Thus predation is also a potentially important biotic interaction. The relative roles of competition and predation in determining the high species richness of such tropical flood-plain rivers are unknown. Habitat selection may be important in reducing interspecific competition in diverse assemblages that are exemplified by the small-bodied fishes living in banks of leaf litter in small, black-water tributaries of the Rio Negro (Henderson and Walker, 1990).

The impact of riverine fishes on other organisms is still unclear (Hildrew, 1990). However, when large predatory fish were excluded from sections of a Californian river, the stands of filamentous green algae increased (Power, 1990). The predators fed on small fish and large invertebrates, which in turn fed on algal-eating chironomid larvae. The exclusion of the large fish meant that the predators of the chironomids could reduce the numbers of the latter and so the stands of algae could develop.

3.5 Biotic interactions and community structure in lakes

Lakes lack the abiotic variability that is a consequence of the variability in flow regimes of rivers. This raises the possibility that biotic interactions play a more important role in determining species richness and relative abundance within lakes than they do within rivers.

Experimental and observational studies on centrarchid assemblages of the littoral zone of lakes in North America suggest that interspecific competition, predation and habitat selection play important roles in structuring such communities (Werner, 1984, 1986; Werner and Mittelbach, 1981). Three sunfish species, the green sunfish (*Lepomis cyanllus*), the bluegill (*L. macrochirus*), and the pumpkinseed (*L. gibbosus*), and one bass, the largemouth (*Micropterus salmoides*), have been the focus of most attention. The first three species feed on invertebrates. The largemouth bass eats invertebrates when small but becomes piscivorous as it grows and includes sunfish in its diet.

The green sunfish is restricted to the shallowest, thickly vegetated part of the littoral, where it feeds on large-bodied invertebrates. Within this restricted habitat, it is competitively superior to the bluegill and pumpkinseed. It is better at exploiting the larger-bodied invertebrates associated with the vegetation and is an aggressive, territorial species that will directly interfere with the two other sunfishes. These occupy deeper vegetated areas and more open water. Large pumpkinseeds can feed on molluscs, whereas the smaller-mouthed bluegill takes zooplankton from the water column in addition to other invertebrates (chapter 5).

When small, the two species have similar diets. Compared with the green sunfish, they are more generalist in their habitat usage. In the absence of the largemouth bass, the bluegill and pumpkinseed will move into less vegetated areas if these provide more profitable foraging areas. In the presence of largemouth bass, the smaller size-classes of sunfish are confined to the more densely vegetated areas where they are at less risk of being eaten by the bass. In this confined habitat and given that the smaller size-classes are more similar in their diet, the risk of predation is lessened at the cost of an increase in the likelihood of competition for food. By restricting the young fish to this habitat, the presence of the predator forces the young through a competitive 'bottleneck'. Both experimental studies and observations on the growth patterns of bluegill and pumpkinseed suggest that such competition occurs.

Some bog lakes in northern Wisconsin have a characteristic assemblage of cyprinid fishes, whereas some have a characteristic assemblage of centrarchids but lack most of the cyprinids found in the other lakes (Rahel, 1984). The cyprinids could tolerate physiologically the abiotic conditions found in both sets of lakes, which suggests that the cyprinids are excluded from the second set by centrarchid predation.

Lakes in which the abiotic conditions are harsh may display biotic interaction between fish species with more starkness than lakes with more benign conditions. Of the bog lakes of Wisconsin which are most acidic, some carry only two species, the mudminnow and the yellow perch (others contain only mudminnows and some lack any fish) (Tonn and Magnuson, 1982; Rahel, 1984). When they are young, the two species have similar diets and the presence of small perch reduces the growth rate of the mudminnow. In turn, the smaller mudminnows are at more risk of predation from older perch. Thus during their ontogeny, as they grow in size the perch are first competitors and then predators of the mudminnow (Tonn et al., 1986).

In benign, more productive lakes, predators will have a wider range of potential prey and so the distribution of any one prey species may be less restricted by the presence of predators. There will also be a wider range of ways of making a living. By a combination of habitat selection and partitioning of food resources, species may reduce the intensity of competition although more potential interactions are possible. This would allow more species to co-exist, although on superficial examination, they seem to be exploiting the same resource base. Habitat selection and the ability to use different types of food are factors that allow the co-existence of high numbers of closely related species as exemplified by the species flocks of cichlids in the African Great Lakes (Fryer and Iles, 1972; Echelle and Kornfield, 1984).

Size-selective predation can also play a more indirect role in structuring the community characteristics of lacustrine fishes. Fish that feed on zooplankton differ in the effectiveness with which they can feed on the size-classes of prey that are available. In Scandinavian lakes, there is evidence that the presence of whitefish (*Coregonus* spp.), in the pelagic region of the lake, restricts Arctic char to the littoral region because the whitefish reduce the size range of zooplankton available to below the size which the char can exploit effectively (Svardson, 1976).

The relative abundances of piscivorous and zooplanktivorous fishes may also influence the density of phytoplankton in a lake (Persson et al., 1988; Vanni et al., 1990). If zooplanktivorous fishes suppress the zooplankton, this could permit the development of higher abundances of phytoplankton when the supply of nutrients and light is not limiting the primary producers. If zooplanktivorous fish are heavily predated by piscivorous fishes, this would allow higher population densities of zooplankton with a consequent reduction in phytoplankton. The overall structure of the lake community will then be the resultant of the effects of the factors that control the rate of photosynthesis (bottom-up control) and the effects of predation (top-down control).

One of the few examples of disease having a major effect on a fish assemblage comes from Windermere in north-west England (Craig, 1987). An epidemic in 1976 wiped out more than 98% of the perch population. Pike, which had previously fed on perch, became more cannibalistic and changed their distribution in the lake.

3.6 Biotic interactions in estuaries

The harsh abiotic conditions of estuaries restrict the number of species found in them and so the potential complexity of the biotic interactions. The species that live in estuaries either thoughout their life or as juveniles, can reach high densities leading to the possibility of intense intraspecific competition.

Field experiments on the mummichog, *Fundulus heteroclitus*, provided evidence that food was a limiting resource and that intraspecific competition occurred (Weisberg and Lotrich, 1986). In Canary Creek, a small estuary on the coast of Delaware (USA), the mummichog is the only species present in spring and summer. When fish were kept in large enclosures at their natural density, they showed the same growth rate as fish in the natural population. However, when the density of the fish in enclosures was half the natural density, the growth was two or three times higher than in the unenclosed population. Fish enclosed at higher than natural densities had significantly reduced growth rates.

The use of estuaries as nursery areas by some marine species suggests that predation rates on small fishes may be lower in estuaries than in other coastal waters. The effectiveness of predators hunting visually for fish prey in the main channel of the estuary will be reduced by the turbidity that often characterizes estuarine waters. Mangrove swamps and sea grass beds provide shelter for juveniles. However, for those species that use the shallow lagoons and pools that fringe estuaries, bird predation can be heavy. About 30% of the sticklebacks that entered salt marsh pools on the edge of the St Lawrence Estuary in Quebec (Canada) were eaten by birds (Whoriskey and FitzGerald, 1985). In the West Kleinemond Estuary in South Africa, the average monthly mortality of juveniles of *Rhabdosargus holubi* (Sparidae) was 30% over a period when the estuary was closed off from the sea preventing emigration (Day *et al.*, 1981). This mortality was mainly due to fish-eating birds including cormorants and herons. In the following year, when the density of *R. holubi* in the estuary was much lower, the average monthly mortality was only 2.5%. Most of the birds had moved to another estuary where fish densities were higher.

3.7   Biotic interactions and community structure in the sea

3.7.1   *Rocky and coral reefs*

In some ways, the fish assemblages found on rocky reefs in temperate waters resemble those of the upper stretches of rivers. The assemblages consist mostly of small-bodied, cryptically-coloured fishes, many of which are adapted to living on or near the bottom and which are able to take refuge both from turbulent water and predators by retreating to holes, crevices and other hiding places. Studies in which all the fish species were removed from a reef, showed that recolonization occurs rapidly and restores both the species richness and the relative abundances that were present prior to the disturbance (Grossman, 1982). This recovery reflects the strong habitat selection shown by many reef fishes (Gibson, 1986). Biotic interactions may also play some role in determining the observed community structure. On a reef in California, the distribution of two species of surf perch, *Embiotoca lateralis* and *E. jacksoni* (Embiotocidae), depended both on habitat selection and interspecific competition (Hixon, 1980). *E. lateralis* dominated the shallow reef areas and fed on invertebrates picked off the algae that covered the rocks. *E. jacksoni* lived in deeper water where it foraged on a 'turf' of sessile invertebrates and small tufts of algae. Field experiments and observations showed that *E. lateralis* was largely excluding *E. jacksoni* from the shallower water, but that the distribution of *E. lateralis* at the study site was a consequence of the habitat selection of that species.

The community structure of coral reef fish assemblages has been the subject of many observational and experimental studies (Sale, 1980, 1988). The profusion and diversity of coral-reef fishes set the scene for an intricate network of predatory, competitive and mutualistic interactions to determine the number of species that share a reef and their relative abundances. The stability of coral reefs over geological time together with their benign abiotic conditions have provided the conditions for the evolution of mechanisms that permit the co-existence of the myriad species. The resources, which are potentially in short supply on a reef, are food, living space and shelter.

One mechanism that will reduce competition for limited resources, a mode of selective segregation, is habitat selection. Species with similar resource requirements do not compete because they occupy different regions of the reef as their preferred habitat (chapter 2).

A second mechanism is resource partitioning, with each potentially competing species specializing on a restricted range of the resources available. Present evidence suggests that carnivores such as the groupers (Serranidae) show resource partitioning of food. This is achieved, at least partly, through size selection of prey, with different species feeding on different-sized prey. Partitioning of food resources is also achieved by morphological differences between related species, which means that prey are obtained in different ways. An example of such morphological differences related to feeding in butterflyfishes (Chaetodontidae) is described in chapter 5.

Another form of resource partitioning seen in carnivorous reef fishes is temporal. Many carnivores are diurnal, feeding during the day and resting at night. A second, smaller group are crepuscular or nocturnal, feeding in fading or dim light.

Species that graze on algae encrusting the surface of the reef such as parrotfish (Scaridae) and damselfish (Pomacentridae) often show considerable overlap in the composition of their diet. Such species may then compete for living space, defending areas of the coral surface on which their food is growing. Such defence can be directed against both conspecifics and other species that have similar diets (chapter 4). Even species that range widely over the reef in search of food may defend living space in the form of a resting site that provides a refuge when the fish is inactive.

The colourfulness of many coral-reef fishes might suggest that predation is a minor factor. However, against the background provided by the coral substrate, some colours may act as camouflage. The holes, caves and crevices of the reef in which fish can hide from predators may also have allowed some species to evolve colour patterns that act as signals in mate recognition or resource defence (Thresher, 1984). There are many predators on or close to a reef but their role in determining patterns of distribution and abundance is still uncertain.

The distribution and abundance of species can depend on mutualistic relationships. Some reef fishes act as cleaner fishes. The wrasse *Labroides* is an example from the Pacific. These eat the ectoparasites infesting the body surface and gills of their clients. The client fishes visit their cleaner fish at well-attended cleaning stations on the reef.

Other mutualistic interactions involve a fish and an invertebrate. Anemonefishes (*Amphiprion* spp.) live in association with sea anemones. The fish gain protection from predators because of the stinging nemato-cysts on the tentacles of the anemone, while the anemone may gain some protection from predators that nip at its tentacles and also obtain scraps of food dropped by the fish. The distribution of the anemonefish is

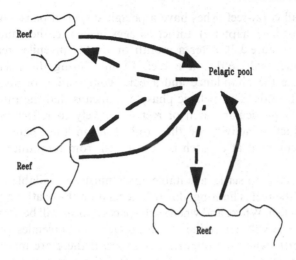

Reefs: Adults and settled juveniles, often space limited

Pelagic: Pool of larvae and pre-settlement juveniles

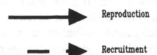

Reproduction

Recruitment

Figure 3.2   Relationship between local reproduction and recruitment for reef fishes with a pelagic disperal phase.

restricted to the distribution of the appropriate anemones; however, the anemone is found in the absence of the fish. There are also associations between gobies and burrowing-shrimps such as that between the goby *Psilogobius* and the shrimp *Alpheus*. The two share a burrow, which protects them both from predation. The shrimp digs the burrow and the goby gives alarm and stimulates the shrimp to retreat back to the burrow if danger threatens.

Not surprisingly, given the complexity of the interactions, the processes that allow the co-existence of so many species within relatively small areas are not completely understood. There is probably a changing balance between chance events associated with the settlement of juvenile fish on the reef and the more predictable outcome of biotic interactions after settlement (Roughgarden, 1986; Sale, 1988).

Almost all coral-reef fishes have a pelagic dispersal phase during their early life history (chapter 6). Either as eggs or larvae, the young become planktonic (Figure 3.2). After a month or so, the juveniles reach a size at which they will settle on a reef. Events during the pelagic phase will influence the abundance and species composition of juveniles that are ready to settle. This pelagic phase also means that the rate of settlement over a particular patch of reef is unlikely to reflect in any way the reproductive activity of the adult fish on that patch. In other words, recruitment at a patch is de-coupled from reproduction at that patch.

For a juvenile to settle, a suitable space must be available on the reef. If space on the reef is limiting, this will depend on the death or emigration of a resident fish. Where and when such spaces appear will be unpredictable and so chance will determine which species have juveniles present and ready to settle when the opportunity arises. If there are many juveniles ready to settle, then post-settlement interactions may determine which ones are finally successful and grow to sexual maturity. If there is a paucity of potential recruits, then which species settle will largely be determined by the chance that has brought a juvenile to the right place at the right time. So long as individuals of a species have sufficiently long life-spans and there are some years in which there is good recruitment to a reef, the species will be present as a member of the reef assemblage.

A consequence of these mechanisms is that on a small area of reef, the species composition and relative abundances will change over time because of the vagaries of recruitment. Over a larger spatial scale, the species composition and relative abundances may be more constant and predictable as the effects of local variations are averaged out and the effects of habitat selection, intra- and interspecific competition and predation are revealed.

In many rivers and lakes, the reproductive activity at a locality directly affects the level of recruitment to that locality (closed-recruitment). The mechanisms that allow species to co-exist in these habitats may differ from those that operate in habitats like coral reefs. In the latter, the species composition at a particular locality depends on a rain of settlers from a pool of potential recruits derived from many, spatially separated localities (open-recruitment) (see also Roughgarden, 1986).

The normal course of events on reefs can be interrupted by catastrophic changes. These may be caused by physical events such as hurricanes (chapter 2), or sudden biotic changes. In the Caribbean in 1983–84, a pathogen killed off a high proportion of the sea urchin *Diadema antillarum*.

This urchin grazed on algal mats on the reef and its disappearance was marked by an increase in the feeding rates of herbivorous fishes.

3.7.2   Open sea

In the open sea, the biotic interactions and the community structure of fish assemblages will reflect the lack of cover that, for many freshwater and littoral fishes, is provided by the substratum and stands of vegetation. The study of the interactions between open-sea fishes is made difficult by the mobility of the fishes and the physical scale of the environment they occupy. The field experiments that have characterized studies on riverine and reef communities are much less feasible. On the other hand, the changes in species richness and the relative abundance of species that are the consequence of intensive commerical fisheries (chapter 7) provide some information. The interpretation of this data is made difficult because the changes are usually an unintended consequence of the fishing, and commercial fishing activities are not designed to provide data on biotic interactions and community structure. Even when changes in species abundance are dramatic, the causes of the changes are often obscure.

Predation is likely to have a major effect on community structure for both pelagic and demersal fishes. Most marine species lay pelagic eggs (chapter 6), so from the moment of fertilization an individual, without parental protection or physical refuge, is at risk of being eaten. This risk declines as the individual grows in size (McGurk, 1986).

When an egg, larva or small juvenile, a fish is at risk from a range of invertebrate predators, which include coelenterates and carnivorous copepods. Such predation can reach high intensities. In a cove in the Gulf of California siphonophores were taking about 28% of the available fish larvae and similar high values have been recorded at other shallow-water localities (Bailey and Houde, 1989). In clupeoids such as the northern anchovy (*Engraulis mordax*), the losses in the early life history stages, through the depradations of cannibalistic larger fish, can form an important component of the total. One estimate suggested that cannibalism accounted for about 32% of egg mortality. Other fish species will also take their toll of these early life history stages.

Not even demersal eggs such as those of the herring are immune from predation. Diving birds feed on the aggregations of eggs laid in shallow waters by the Pacific herring, although the main losses may be due to predation by small invertebrates including snails and amphipods (Bailey and Houde, 1989).

As they grow, the fishes in coastal waters are at risk from fish-eating birds that form often dense nesting colonies at coastal sites. In many parts of the world, these bird predators include cormorants, gulls, auks, pelicans and gannets. Estimates based on the food requirements of individual birds and population densities suggest that such predation can account for considerable portions of the annual fish production in those areas exploited by the birds (Furness, 1982). Birds may take over 20% of the production in areas such as the coastal waters off Oregon (USA), the areas around the Shetland Islands off Scotland, and off the south-western coast of Africa. These rates of exploitation can mean that the birds are, at times, competing with commercial fisheries for the fish (see also chapter 7). This is more likely when the fishery is for relatively small-bodied fishes such as anchovies, pilchards and sand-eels.

With further growth, fishes become susceptible to a narrower and narrower range of predators, which include larger, fish-eating mammals and the largest piscivorous fishes including sharks. The decline in the mortality rate with an increase in body size of fishes reflects this decreasing risk of predation. However, there is little understanding of how this size-dependent predation influences the species composition and relative abundances of the fish assemblages.

Body coloration reflects the importance of predation. Epipelagic fishes show countershading with a darker back and a silvery belly, which presumably makes them less visible to predators in the downwelling light. In some pelagic species the flanks are silvery and, by reflecting light, make the fish less visible. Benthic fishes have a drab, brownish, dorsal coloration that blends with the substratum. Some have the ability to modify their pigmentation to match the colour of the background (Lythgoe, 1979).

The effects of predation on the assemblages of marine fishes are likely to interact with the effects of inter- and intraspecific competition. Direct evidence of competition is difficult to obtain in populations of marine species. Some evidence is provided if there are improvements in the growth rate, reproductive rate or survival when commercial fishing activities cause large reductions in the abundances of populations of potential competitors. An increase in the growth rate of the gadoids, haddock and whiting, was observed when abundances were reduced and both species showed a reduction in the age of maturity (Furness, 1982). The problem with such observations is that other important factors including changing abiotic conditions may be occurring at the same time as the populations are being reduced by fishing. Consequently, the changes observed cannot be unambiguously ascribed to a reduction in the intensity of competition, presumably for food.

Georges Bank is a shallow area off the east coast of Massachusetts (USA) and the scene of major fishing activity. In the period 1962–1971, the abundance of haddock was reduced by about 90%. As the haddock declined in abundance, other members of the demersal assemblage increased in abundance including winter skate (*Raja ocellata*), little skate (*Raja erinacea*) and the window pane (*Scophthalmus aquosus*). An attempt was made to reconstruct these events by developing a model of the Georges Bank fishes, which assumed that the changes were due to biotic interactions rather than abiotic changes. Both interspecific competition and predation were assumed to be relevant interactions between the juveniles of the main species present (Overholtz and Tyler, 1986). Although the model had some success, it failed to reproduce some of the important changes that took place. This relative failure indicates that, even for a well-studied shelf assemblage, the consequences of biotic interactions are poorly understood.

Disease can have sporadic effects on inshore fish assemblages. In the 1950s, the densities of both herring and mackerel in the Gulf of St Lawrence were reduced to low levels because of a fungus infection. Subsequently, both pelagic species recovered but the pattern of recovery suggested that both intra- and interspecific competition, cannibalism and predation influenced the abundances of the two species (Winters, 1976).

Although the feeding interactions of epipelagic assemblages of the open ocean are partly known (chapter 5), the consequences of predation and competition can only be inferred indirectly. A striking pattern is the vertical migration of many mesopelagic species (chapter 4), which allows them to exploit the richer feeding in the surface layers during the night, while avoiding interactions with the piscivorous fishes that hunt primarily during the day-time. There is a rapid decline in the amount of food available to fishes with an increase in depth (Marshall, 1979), so the night-time migration to the surface may reduce competition between mesopelagic species for food.

In the non-migrating mesopelagic species of which the stomiatoid *Cyclothone* is the most prominent, there is evidence of a vertical partitioning of the habitat (Marshall, 1979), with different species living over different ranges of depth. A consequence of this may be a reduction in the intensity of competition for what food is available. The non-migrating mesopelagic species tend to have relatively bigger gapes compared with the migrators, which should allow them to exploit a wider size-range of prey. Many of the small mesopelagic species, including the lantern fishes, are bioluminescent. One arrangement of the light-emitting patches on the body reduces the contrast between the body and its surroundings to a potential predator that is looking upwards towards the surface.

The few species that are found in bathypelagic zone are living in a cold, dark and extremely impoverished habitat. They have morphological adaptations that allow them to feed on large-bodied prey. The abiotic conditions are almost uniform over vast volumes of water, but of the effects of biotic interactions on the species richness and relative abundances of these deep-sea assemblages virtually nothing is known.

CHAPTER FOUR

MIGRATION, TERRITORIALITY AND SHOALING IN FISHES

4.1 Introduction

Water movements range from the torrential discharges of mountain streams, the crash of waves against a rocky shore, the ebb and flow of the tide over a gently shelving beach to the slow moving currents of the oceans. Unless fish show active behavioural mechanisms to prevent it, they will be carried along with the water flow. Eggs, larvae and young juveniles either lack the mechanisms for active movement or those mechanisms are too poorly developed to prevent passive displacement. In these early stages in the life history, the movements of fish are largely determined by when and where the eggs are laid (chapter 6). As the locomotory capacity matures, fish become capable of resisting the water movements and can make active choices about the habitat in which they will live (chapters 2 and 3).

These choices can be described in terms of two dimensions: (i) the attachment shown by individual fish to a given site; and (ii) the nature of the social interactions with other fishes of the same species (conspecific interactions) and with other species (heterospecific interactions) (Table 4.1). In many species, individual fish will change their behaviour during their lifetime and so will change where they can be classified with respect to these two dimensions. Before exploring this classification in more detail, it will be useful to consider the swimming capacity of fishes in relation to their size and the energy costs of swimming at different speeds.

4.2 Swimming capacity and energy costs

The correlations between fish shape and habitat are described in chapter 2. Size is the other major factor influencing swimming capacity. Size affects

Table 4.1 Classification of pattern of movement based on two dimensions: site attachment and nature of social interaction

Social behaviour	Site attachment		
	Strong	Weak	None
Hostile	Territoriality	Weak territoriality	Inter-individual distance
Neutral	Home range	Weak home range/ wandering	Wandering
Gregarious	Shoal in home range	Shoal in weak home range/wandering	Wandering shoal

swimming in two ways. It determines the hydrodynamic regime in which the fish operates (chapter 1) and the absolute velocity that can be achieved.

Larval teleosts experience viscous drag as a major force in their hydrodynamic environment (chapter 1). Its effect is that movement stops as soon as locomotion stops, the larval fish is unable to glide forward without actively swimming (Webb and Weihs, 1986). For larger fish, viscous drag becomes less important, while pressure drag becomes dominant (Table 1.2). In this hydrodynamic regime, which is essentially occupied by most juvenile and all adult fishes, the animal can glide forward though the water at the end of a period of active propulsion. Such burst and coast swimming is characterized by a one or more locomotory cycles of the body and fins, which accelerate the fish (burst) followed by a phase of gliding with no propulsive movements. The fish gains an advantage because during the glide phase it can maintain its body in a straight line, which minimizes drag. During the propulsive phase, the body must be flexed and this increases the drag.

When discussing swimming speeds, body size and the energy costs of swimming, it is useful to define three ranges of speeds: sustained, prolonged and burst swimming (Hoar and Randall, 1978). In sustained swimming, the energy required by the swimming muscles is provided by aerobic respiration. The fish does not build up an oxygen debt and fatigues only slowly. Swimming velocities are usually restricted to less than six to seven body lengths per second, and typically much lower speeds are maintained during long swimming bouts. In burst swimming, much higher speeds can be reached, sometimes as high as 20 body lengths per second. However, the power for burst swimming is generated by anaerobic respiration and fatigue is extremely rapid. Prolonged swimming is an intermediate category covering speeds between those typical of burst and sustained swimming. Energy is supplied by both aerobic and anaerobic metabolism. Bouts of prolonged swimming may last from about 15 to 3 h but long bouts end in fatigue. The capacity to show short bursts of high-speed swimming can be crucial for fish chasing prey, escaping from predators (or fishing nets) or making their way through fast-flowing waters. The ability to sustain swimming will be important for foraging over large areas, holding position in the face of slow but steady currents or migrating long distances. For comparisons of fishes of different sizes or species, critical swimming speed is often used. It is the maximum speed achieved before fatigue, when the fish is subjected to stepwise increases in swimming speed in a swimming chamber.

The maximum velocity increases with the size of the fish. This increase can often be described by the relationship:

$$\log V = a + b \log L$$

where V is swimming velocity in mm s^{-1} and L is fish body length in mm (Beamish, 1978). For example, a sockeye salmon 100 mm in length has a critical speed of about 600 mm s^{-1} (6 body lengths per second) but at 500 mm its critical speed is about 1400 mm s^{-1}. When swimming speed is measured in body lengths rather than millimetres per second, smaller fish tend to perform better than larger fish although their absolute speeds are lower.

Swimming is an energetically costly form of locomotion (Schmidt–Nielson, 1984). One measure of this cost is the energy expended in moving a gram of fish one kilometre (J g^{-1} km^{-1}). This cost decreases with an increase in the size of the fish, but for those species studied so far, the cost is similar for fish of a comparable size. For a sockeye salmon weighing 10 g, the energy cost of travelling a kilometre is about 46 joules (4.6 J g^{-1} km^{-1}) but for a sockeye weighing 1000 g the cost is about 1430 joules (1.4 J g^{-1} km^{-1}) (Beamish, 1978; Schmidt Nielson, 1984). Larger fish may be more likely to undertake long-distance migrations than smaller fish because the former travel more efficiently than the latter (Roff, 1988). These values refer to sustained swimming. Burst swimming is much more costly. For juvenile coho salmon, burst swimming can be nearly 40 times more costly in energy terms than swimming at the maximum sustainable speed (Puckett and Dill, 1984). An ability to escape from the lunge of a predator is a matter of life or death; speed of movement will be more important than efficiency of movement.

4.3 Patterns of site attachment and social interactions

4.3.1 *Home range and territoriality*

Anglers sometimes 'spin' tales of special fish that lurk in known localities. Such observations refer to the concept of site attachment. An individual lives all or part of its life in an area that is much smaller than the area potentially available given the swimming capacity of the fish. Evidence for site attachment in fishes was reviewed by Gerking (1959).

In some species, individual fish show site attachment but the areas occupied by different individuals overlap and there are no behavioural mechanisms by which an individual excludes another from a given area. Each individual occupies a home range. Residence within a home range probably allows the fish to learn the sites of food and shelter and in some cases, suitable sites for reproduction. If the resident (or residents) in an area can exclude other individuals from the area, then the resident is displaying territoriality. Territorial behaviour allows the occupants to sequester for themselves the resources of that area. The benefits of such sequestration must be offset against any costs that have to be met to maintain the exclusion of other fishes. The area may be defended only against conspecifics or in some examples against both con- and heterospecifics. The resource sequestered may be food, shelter, spawning sites or combinations of these.

4.3.2 *Shoaling*

For part of all of their life, fish may spend their time associated with other fish in shoals. The formation of a shoal depends on the fish reacting to each other socially and not just on the fish coming together at sites that are attractive to them for some reason not related to the presence of other fish. The terms shoal and school are sometimes used interchangeably, but it is useful to retain school for the situation in which the fish in a shoal show a high degree of co-ordination in their spatial positions within the shoal and well-synchronized manoeuvres (Pitcher, 1986).

A fish in a shoal or school probably has a reduced risk of being attacked by a predator (Pitcher, 1986). It may also improve the individual's chances of finding food, which is dispersed in patches through the habitat. A likely cost of being in a shoal is increased competition for food when it is found because membership of a shoal ensures that the individual is surrounded by other individuals.

In some species, shoals may range over vast areas, whereas in others the shoal may occupy a home range. The use of space by individual shoals is a topic on which more information is urgently required.

4.4 Migration

4.4.1 *Definitions of migration*

Many shoaling species are also migratory. A useful definition of migration is those movements that result in an alternation between two more separate habitats and which occur with a regular periodicity and involve a large proportion of the population (Northcote, 1978). The fish come and go between habitats within a lifetime, either once or periodically (Roff, 1988). For migration to take place, the fish must lose any site attachment they may have had, and which they may assume again at the termination of the migratory phase. As measured by population abundance, many migratory species are highly successful, which indicates that, in terms of reproductive success, the benefits of the migratory behaviour outweigh the inevitable costs. The costs will include the energy costs of the movements and the possibility of increased predation because the fish are moving through unfamiliar localities. The benefits accrue because the movements bring the fish into habitats that are more suitable for them than the habitats they have left.

Typically, the migrations take fish between those habitats that are suitable for feeding and habitats that are suitable for reproduction (Figure 4.1). They may also include movements to and from habitats that form refuges from environments that are unfavourable either because of adverse abiotic conditions (chapter 2) or because the presence of predators or pathogens make them dangerous (chapter 3). A nomenclature for describing fish migration has now been developed (Harden Jones, 1968; McDowall, 1988).

Potamodromous species undergo migrations that are entirely restricted to fresh waters. They may involve migrations within river systems or migrations between rivers and lakes. *Oceanodromous* species migrate entirely within ocean systems. Many of the most important commercially exploited species are oceanodromous. *Diadromous* species migrate between marine and freshwater systems (Figure 4.2). Three categories of diadromy are recognized. *Anadromous* species migrate into fresh water to spawn but spend a portion of their life feeding in the sea. *Catadromous* species migrate

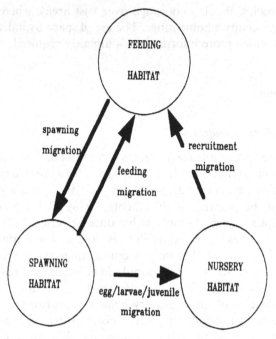

Figure 4.1 Basic pattern of migration between feeding, spawning and nursery habitats.

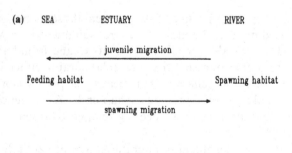

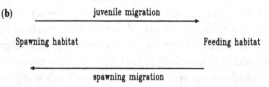

Figure 4.2 Anadromous (a) and catadromous (b) migration.

to the sea to spawn but spend a portion of their life-cycle feeding in fresh water. *Amphidromous* species migrate between sea and fresh water but the migrations are not directly linked to reproduction. The characteristic of diadromous species is that they have the physiological capacity to move between waters with very different osmotic and ionic properties (chapter 1).

4.5 Fish movements in rivers

4.5.1 *Home range and territoriality*

Rivers present two features of importance for fish movement. The first is that the flow is essentially unidirectional so upstream movement or even simply holding station in the current will depend on the active swimming of the fish. In contrast, downstream movement can take place passively with the fish drifting with the current. In many rivers, high waterfalls or rapids act as barriers to upstream movement. In the stretches above the barrier, the populations will only persist if the fish show behavioural traits that reduce the downstream losses to a level that permits the continued existence of the population (Northcote, 1978). The banks and bottom of a river provide a relatively stable topography that fish can use, if necessary, to orientate in relation to particular sites.

In the upper stretches of northern temperate rivers, the salmonids provide a good example of social organization in relation to the physical environment. Many salmonids species lay their eggs in gravel in small fast-flowing streams. Although in some anadromous salmonids, the young move down to the sea soon after they emerge from the gravel, in other species there is a longer period of stream residency before migration (in some cases, the whole life-cycle takes place in fresh water). The fish take up stations that allow them access to prey drifting past but also provide shelter from the current. Stations are defended as territories. However, a fish will move from station to station, and as it leaves one station it will be re-occupied by another fish. Some stations are preferred to others and access to these stations depends on the dominance hierarchy that is established between the fish in a given locality (Jenkins, 1969). This social system is called partial territoriality. In addition to resident fish, there are also more mobile individuals that move from station to station as they become available but show little or no permanent site attachment. The preferred sites may be those at which the difference between the energy that can be gained from the drifting food and the energy cost of maintaining station is highest (Fausch, 1984). These sites may become loci for interspecific competition between salmonid species (chapter 3).

The current flowing past a site ensures that the supply of drift is renewed continually. If the resource that is being exploited is not renewed rapidly, it may not be economical to defend a site. Species feeding on invertebrates living in the substrate or on highly mobile prey often range widely as they forage. The fishes may live in shoals or forage as individuals. Even salmonids, as they become large and include fish in their diet (chapter 5), are more mobile and less likely to defend a feeding station. Evidence from marked fish suggests that shoals of cyprinids like roach may occupy a home range; however, the same studies also suggest that a portion of the population is mobile and not restricted to a home range (Stott, 1967). This mobile portion of the population may consist of individuals that move from one adjacent shoal to the next and so shift their home range. It may also consist of individuals that have no tendency to stay within a home range.

4.5.2 *Migration in rivers*

Migrations in rivers are usually associated with either reproduction, the opening up of feeding grounds as a result of seasonal changes in water levels or movement to over-wintering areas.

The upper stretches of rivers are usually characterized by well-oxygenated water and a current sufficently high to prevent the settlement of silt. The density of predators on fish eggs and larvae may also be low. This combination of factors probably accounts for the tendency of many riverine species to migrate upstream for spawning. The salmonids provide a good example of this but other freshwater fishes show the same tendency. In larger river systems, these upstream migrations cover large distances. Large characins living in the Paraguai River system of South America migrate over 400 km from their feeding grounds in the Lower Chaco to their breeding sites in the Andean foothills (Lowe-McConnell, 1987).

In flood plain rivers, seasonal high waters spill over the river banks and flood the surrounding land, opening up large feeding areas for the river fishes. The pattern of movement of large characins in the Rio Madeira in the Amazon system illustrates how this flooded land is exploited (Goulding, 1980). Early in the flood, the fish move down the tributaries of the Rio Madeira to spawn in the main river channel. When spent, they move back up the tributaries and spread out over the forest floor to feed (chapter 5). As the waters begin to fall, the fish move off the forest floor into the tributaries and down to the main river. During the low water season, they move upstream intermittently, before moving again into tributaries further upstream as water levels rise. As the fish get older, they move further and further upstream in the river system, while the population downstream is replenished by the eggs and larvae, which are swept downstream in the main channel. The young move into sloughs, flood plain lakes and the flooded forest to grow rapidly though the small and vulnerable size classes.

A feature of most river systems is this movement of the young fish into shallow sheltered waters where they are less likely to be swept away and have a good food supply. In the management of a river system, the provision of such areas will be of vital importance for the maintenance of many fish populations (Mills and Mann, 1985).

In the rivers of temperate and sub-polar regions, large juveniles and adults may migrate to over-wintering areas. In the low-lying fenland of eastern England in winter, high densities of fish were found in boatyards linked to the rivers, where the pleasure boats were moored (Jordan and Wortley, 1985). The boats provided shelter equivalent to that given by natural over-hanging banks or tree stumps.

Some freshwater species migrate between rivers and lakes. Freshwater salmonids including the brown and rainbow trout will spend one or more years living in a stream maintaining partial territories and feeding on drift. However, then the fish migrate down (or up) to a lake in the river system. In the lake, they range much more widely for they can no longer rely on a continually replenished supply of drift. On becoming sexually mature, they re-ascend a stream to spawn, usually choosing the stream in which they hatched and spent their first years of life. A similar upstream spawning migration is seen in some tropical lake-dwelling species including cyprinids (Lowe-McConnell, 1987).

4.6 Social structure and movement in lakes

In lakes, the social structure and movement of fishes is related not to water currents but to the presence of a littoral zone and a pelagic zone. The former resembles a river in that fixed topographical features are present. The latter resembles the open sea in its lack of obvious physical structure.

4.6.1 *Social structure*

Species that live in the littoral zone may be strongly territorial. An example is the green sunfish, which is essentially confined to shallow, vegetated littoral areas in eastern North America (Gerking, 1959). Other littoral species probably live in home ranges, for example lake brown trout, but these ranges may be poorly defined spatially as in the northern pike. In some species, the fish live essentially solitary lives like the pike. In other species the fish live in larger or smaller shoals, for example the shiners, *Notropis* spp., cyprinids of North American lakes. A species like the yellow perch lives in shoals when young, but with increasing age and size, the fish become more and more solitary (Helfman, 1981).

In large lakes, the littoral zone may be broken up into several different habitats occupied by different species. For example, the western shoreline of Lake Malawi consists of alternating patches of sand, weed and rock (McKaye and Gray, 1984). Many of the cichlid species that occupy the rocky habitats (the 'Mbuna' species) show strong site attachment, staying within a few square metres all their life. This attachment may be one of the reasons for the many species of cichlid found in Lake Malawi because this reluctance to move would ensure little gene flow between populations occupying different rocky shores (Fryer and Iles, 1972; Lowe-McConnell, 1987). However, some 'Mbuna' are not totally sedentary. Artificial rocky reefs sited in sandy areas 1 km from the nearest natural reef were colonized by some species of 'Mbuna' (McKaye and Gray, 1984). This recalls the observation that riverine populations include both resident and mobile components.

Many pelagic lake species like the clupeids and whitefishes live in shoals, and this mode of social organization in a physically unbounded environment will be met again in the pelagic species of the seas.

4.6.2 *Migration in lakes*

The migrations into inflowing or outflowing streams shown by some lake-dwelling fishes has already been mentioned. Other lacustrine species migrate to spawning grounds. For example, in Lake Mendota in the USA, white bass, *Morone chrysops*, migrate from the epipelagic zone of the lake to well-defined spawning areas on the edge of the lake.

A feature of some populations living in the pelagic zone is a vertical migration, with the fish moving up and down in the water column on a daily basis. These migrations typically involve species feeding on zooplankton such as clupeoid species in Lake Volta in west Africa (Lowe-McConnell, 1987). Two migratory patterns have been recognized (Neilson and Perry, 1990). In Type I migration, the fish ascend at dusk and descend around dawn. In Type II migration, the descent is at dusk and the ascent at dawn.

An example of a Type I migration is that of sockeye salmon (*Oncorhynchus nerka*) fry in Babine Lake in western Canada (Brett, 1983). These fry live in shoals feeding on zooplankton. In the summer, when the lake is stratified, the fish spend the day in cool deep water. At dusk they migrate into the warmer surface water to feed, dropping into slightly deeper water as they become satiated. Just before dawn they again feed in the surface waters before dropping back into the deep cool water. A possible function for this vertical migration is that it minimizes the risk of predation to the fry while still allowing them time to feed in the zooplankton-rich surface waters (Clark and Levy, 1988). A second, but not mutually exclusive explanation, is that the fry gain a bioenergetic advantage by feeding in warm waters where their metabolic rate is relatively high but resting with a lower metabolic rate in cool waters (Brett, 1983). This latter explanation has been made more unlikely by a recent experiment on bluegill sunfish (Wildhaber and Crowder, 1990). The bluegill failed to chose combinations of temperatures and food levels that would have maximized growth. Juvenile roach in lakes show both Types I and II migrations.

4.7 Social structure and movement on reefs and rocky shores

The physical complexity of reefs provides cues, which, at least potentially, provide the fishes with means of finding their shelters again reliably. Many reef fishes show strong site attachment throughout some or all of their life on the reef.

In species like the sculpin, *Oligocottus maculosus*, which lives on rocky shores of western North America, the fish occupy a home range but do not defend its borders. A sculpin will occupy the same rock pool for several weeks or months (Green, 1971). Other species are strongly territorial. On coral reefs, territories may be defended not only against conspecifics but also against other species (Sale, 1980). Such interspecific territorial defence is sometimes most strongly directed against species that have similar diets to the territory holder (Myrberg and Thresher, 1974). Other species form shoals and these shoals allow the fish to invade territories because the territory owner is swamped by numbers, for example shoals of surgeonfish and parrotfish will invade the feeding territories of damselfish (Sale, 1980). Other shoaling reef fishes exploit zooplankton above the reef. Species like the grunts (*Haemulon* spp.) migrate from the reef following well-defined routes to feed in sea-grass beds returning to the reef to shelter.

Species like some groupers migrate to the outer edge of the reef to spawn, the fish forming large spawning aggregations at the spawning sites (chapter 6).

A feature of coral-reef fishes, which is seen to a lesser extent in other shallow water fish faunas, is a turnover at dawn and dusk between an assemblage of fishes active during the day (diurnal) and an assemblage active at night (nocturnal) (Helfman, 1986). This pattern produces a segregation in time between fishes that may be exploiting similar prey.

Many reef species have the ability to home to their territory or home range, so if they are displaced or have taken advantage of high water to move away from their usual living area, they can make their way back. *O. maculosus*, a fish typically less than about 150 mm long, could find its way back to a home pool after being displaced as much as 90 m (Green, 1971). The goby, *Bathygobius soporator*, can leap accurately from one pool to another at low water, to escape from a disturbance (Aronson, 1951).

When territorial fish from an area of reef are removed their places are often quickly taken by others (Sale, 1980). This suggests that at least portions of the populations of reef-dwelling fishes are mobile and ready to move in to vacant sites as they become available.

4.8 Social structure and movements of marine fishes

4.8.1 *Pelagic fishes*

The fishes of open waters may be solitary, but in many species, individual fish live out all or most of their lives in schools. Good examples of schooling pelagic fishes are provided by the planktivorous herrings, anchovies and other clupeoid fishes, the carnivorous scombrid fishes such as mackerel, and the fish- and squid-eating tunas. Pelagic fishes live in an environment that offers them no holes or crevices in which to shelter. Their food is distributed in mobile patches of phyto- or zooplankton or is the species exploiting such patches. In this environment, there is little or no possibility of site attachment, although fish will collect underneath floating debris or will associate with structures such as off-shore drilling rigs.

The dynamics of movement of the schools of pelagic fishes are not well known. Do schools move at random or occupy a diffuse, ill-defined home range? How frequently do individual fish switch between schools or is membership of a school for life? Does the size of a school change in relation to environmental factors such as food availability?

4.8.2 Migrations of pelagic fishes

During their life-span, many pelagic fishes make oceanodromous migrations that cover many hundreds or thousands of kilometres. Most is now known about the migrations of the pelagic fishes of the continental shelf that are or have been the prey of commercial fisheries. At the height of the herring fishery in the North Sea, the migrations of the fish were mirrored by the migrations up and down the east coast of Scotland and northern England of the fishing boats and the shore workers who gutted the fish.

There are several populations of herring in the North Sea that spawn on different gravel banks (Figure 4.3) (Harden Jones, 1968). The Bank population has spawning grounds on the Dogger Bank and off the east coast of England to the west of the Bank. The fish move in late summer from feeding grounds in the northern North Sea southwards to the spawning grounds where they spawn in the autumn. The spent fish migrate north westwards to wintering grounds in the Skagerrak between Denmark and Norway. Then, in early spring, they move north west to the feeding grounds, thus completing an anti-clockwise circuit. This pattern of movement correlates with the pattern of surface-water currents in the North Sea. The larval herring are carried inshore, with sheltered shores, bays and estuaries acting as nursery areas in a way that parallels the use of backwaters by the young of riverine fishes.

The migrations of the pelagic fishes of the open oceans are less well-known. Commercially important tuna species show extensive movements (Figure 4.4). Northern bluefin tuna tagged in the eastern Pacific off the coast of North America have been recaptured off Japan, and other tagged tuna have made the trans-Pacific journey in the other direction (Bayliff, 1980).

4.8.3 Vertical migrations of pelagic fishes

Some pelagic species such as the herring show vertical migrations comparable with those shown by pelagic fishes in lakes. During the night

Figure 4.3 Migrations of North Sea herring (*Clupea harengus*) stocks: spawning grounds in black; feeding grounds, vertical hatching; overwintering grounds, cross hatching; spawning migration, fine arrows; feeding migration, thick arrows. Redrawn from F.A.O. (1981).

the herring move into the surface waters, but they spend the day in deeper waters or close to the bottom. Many of the species of lantern fish (Myctophidae) spend the day at depths of several hundred metres where there is little or no sunlight. At night they migrate to the surface layers, feeding on zooplankton at depths of 20 to 200 metres. Not all mesopelagic fishes show vertical migration of such large amplitude and some stay within the same depth range throughout the daily cycle (Marshall, 1979; Longhurst and Pauly, 1987).

An understanding of the adaptive significance of such vertical migrations will come from analyses of the benefits and costs of such movements (Clark and Levy, 1988; Neilsen and Perry, 1990).

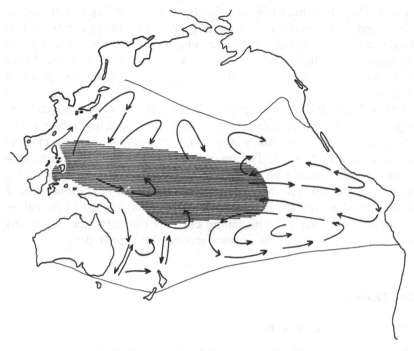

Figure 4.4 Migrations of skipjack tuna (*Katsuwonus pelamis*) in Pacific Ocean: spawning areas, horizontal hatching; northern and southern distribution limits, thin line. Redrawn from F.A.O. (1981).

4.8.4 *Social structure and movements of demersal fishes*

The difficulties of observing the fish living on or close to the bottom of the sea mean that there is little detailed knowledge of the social structure of such species. Movements of tagged plaice suggest that there is no territoriality or home-range behaviour in these flatfish. In the southern North Sea, the pelagic, larval plaice are carried by currents towards the Dutch coast. At metamorphosis, the juveniles settle on sheltered, shallow, sandy shores. As they grow they move into deeper water. When adult, the plaice migrate to the south to spawn, with the spent fish returning north to the feeding grounds (Harden Jones, 1968). During movement, the plaice use tidal currents by a behavioural process called selective tidal stream transport (Arnold, 1981). Towards the end of the slack water period of the tidal cycle, the fish leave the bottom and remain in mid-water for several hours, returning to the bottom at the next slack water. In mid-water, the fish tend to be carried downstream by the prevailing tidal current. The fish swim actively but slowly, rarely exceeding 1 body lengths per second. The direction of swimming does not always correspond to the direction of the current. Estimates suggest that, compared with a fish not using the method, a plaice using selective tidal transport could save up to 40% of the energy costs of a migration to or from the spawning grounds.

Other demersal fishes exemplified by the cod also show long-distance migrations between spawning, feeding and over-wintering habitats. The Arctic cod migrates from the waters north of Norway to spawn close to the west coast of Norway. The pelagic eggs and larvae drift northwards in the coastal waters, with the growing young fish moving close to the bottom to take up their demersal life. The spawning migrations of both demersal and pelagic fishes result in the eggs being deposited in regions of the sea from which the prevailing currents will carry the developing eggs and larvae towards the nursery grounds by passive drift.

4.9 Diadromy

4.9.1 *Distribution of diadromy*

Those species in which all or some fish move on a regular basis between fresh and marine waters include some of the best-known migratory species including the Pacific salmon of the north Pacific, the European and American eels and the hilsa (*Hilsa ilisha*) of India and south east Asia. Diadromous fishes include species of the primitive but specialized jawless lampreys, and this mode of life is well represented in the primitive teleost orders, the Clupeiformes, Anguilliformes and Salmoniformes. It is much more sporadically represented in evolutionary more-advanced teleosts (McDowall, 1988).

There are geographical patterns in the distribution of anadromous and catadromous fishes (Gross, 1987; McDowall, 1988). The frequency of anadromous species is greatest between 30° and 60° N (Figure 4.5). This peak reflects the geographical distribution of lampreys and salmonids. Catadromous species are found between about 60° N and 50° S with a peak in the tropics and sub-tropics (Figure 4.6). This reflects the geographical distribution of the eels (Anguillidae) and the mullets (Mugilidae).

If expressed as the proportion of diadromous species, the geographical pattern is even clearer. The proportion of anadromous species increases

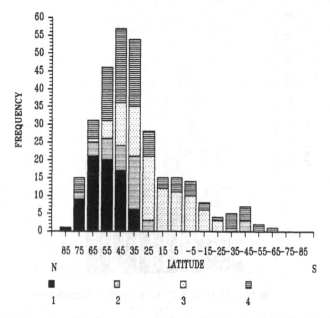

Figure 4.5 Frequency of anadromous species in relation to latitude. Northerly latitudes are shown as positive values and southerly latitudes as negative values. Key: (1) Salmonidae; (2) other Salmoniformes; (3) Clupeiformes; (4) other fishes. Data from McDowell (1988).

with latitude, the proportion of catadromous species decreases with latitude. Two explanations, not mutually exclusive, may account for these distributions. The pattern may be a reflection of the phylogenetic correlations of these types of migration. The salmonids are cold-water fishes and also happen to be anadromous, while the eels are essentially a group from warm waters and also happen to be catadromous. On this explanation, the correlation between latitude and type of migration is coincidental.

A more interesting possibility is that the pattern reflects the relative productivity of fresh waters and sea waters at different latitudes (Gross, 1987). This hypothesis suggests that, at high latitudes, the productivity of the sea is higher than that of fresh waters. Fish that migrate from fresh water to the sea move to richer feeding grounds and consequently show increased growth compared with fish that stay in fresh water. At low latitudes, the fresh waters tend to be more productive than the seas and the move to richer feeding grounds takes the fish from sea water into fresh water. Any migrations in relation to productivity gradients are likely to

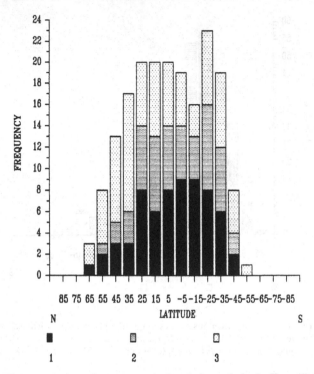

Figure 4.6   Frequency of catadromous species in relation to latitude. Key : (1) Anguillidae; (2) Mugilidae; (3) other fishes. Data from McDowell (1988).

be modified by factors such as the need to avoid extremely cold sea waters or periods of drought when freshwater habitats are likely to dry up.

4.9.2   *Social structure of diadromous fishes*

The range of social structures that may be shown within a single life-span is well illustrated by the anadromous salmonids such as the Atlantic salmon or the sockeye salmon of the northern Pacific (Mills, 1981; Foester, 1968). On emergence from the gravel redd in which the eggs are spawned, the young Atlantic salmon takes up a feeding territory, which provides it with a series of stations from which it can feed. After a year or more, it undergoes the transformation to a smolt and moves downstream towards the sea, the smolts moving downstream in shoals. Once in the sea, the fish migrate to feeding grounds off the coast of Greenland. Eventually, the maturing adults migrate back into fresh water, usually homing to the stream in which they were born.

On the breeding grounds, the males defend territories within which the females dig the redds in which the eggs are laid. While on these breeding territories, the adults do not feed. The general pattern is similar for the sockeye, except that after hatching, the young fish (fry) move downstream into a lake. Within the lake they live in shoals feeding on zooplankton. After usually a year in the lake, they smolt and migrate to the sea in shoals. They move through the northern Pacific, essentially following the great anti-clockwise North Pacific gyre. With approaching maturity they return to fresh water. Again, on the spawning ground the males defend territories within which the females spawn.

In both Atlantic salmon and sockeye salmon, some fish do not migrate but live out their whole life within fresh water. In Atlantic salmon, these are typically males that become sexually mature at a young age and small size—the precocial males. In sockeye salmon, both females and males may stay in fresh water. They are called Kokanee.

4.10 Homing

There is good evidence that many migratory species return to or close to the place where they were born in order to spawn. In some migratory salmonid populations, fewer than 5% of the returning fish fail to return to their natal stream (Quinn, 1984). In the herring and plaice of the North Sea, the adults from different populations may mix to a greater or lesser extent on the feeding grounds, but their spawning migration takes them back to their natal spawning grounds. How do fish find their way back to the breeding ground on which they were spawned or back to their home pool?

4.10.1 Mechanisms of homing

The environment provides many cues which could, in theory, be used by homing fish (Smith, 1985). These include water currents, gradients in temperature and salinity, olfactory stimuli, the position of the sun during the day or the stars at night, the pattern of polarized light in the sky during the day-time and the variations in the earth's magnetic field. This plethora of potential cues poses two questions: which of them *can* be used and which of them *are* used during a migration?

Experimental studies suggest that some migratory species could, in principle, use several of these cues, but the question still remains—do they use them? The salmonids have provided much of the evidence (and much of the controversy).

There is general agreement that once the migrating, adult salmon has reached fresh water, it relies on olfactory cues to find its way to the appropriate spawning grounds. Salmon that have had their nostrils plugged or the olfactory nerve damaged are unable to discriminate between 'home' water and other water (Hasler and Scholz, 1983). There is controversy about the nature of the olfactory cues used. One hypothesis suggests that the young fish learns the olfactory characteristics of its home stream during its juvenile residency in fresh water. It retains this memory during its period of feeding in the sea and as a returning adult it swims upstream following the odour of its home stream (Hasler and Scholz, 1983). An alternative hypothesis suggests that the olfactory cue is provided by pheromones released by juvenile fish in the home stream and the returning adult fish has an innate ability to respond to the pheromones that are specific to its population (Stabell, 1984). It is even possible that the fish use a combination of both mechanisms making full use of all the olfactory cues that are available.

There is much more uncertainty about how the fish find the mouth of their home river after long sea migrations. One possibility is that the fish show slightly directional movements that take them to the appropriate coastline (biased random search) and then search the coastline for the plume of fresh water carrying the appropriate home-water odours. At the other extreme is the possibility that the fish have a well-developed navigational ability, which allows them to home directly and with high accuracy. The navigational ability is assumed to be based on the detection of the position and change in position of the sun, and in overcast conditions, on the earth's magnetic field. Other studies have suggested that even during the oceanic phase of their migration, the salmon are using olfactory cues in association with currents and gradients in salinity and temperature. Quinn (1984) summarized the evidence that the oceanic phase of the migration back to fresh water of Pacific salmon was a directed rather than a biased random movement. The evidence includes the observations that: (i) fish tagged at a given location in the sea disperse to a wide range of home streams; (ii) tagging studies indicate a relatively rapid movement to precise coastal areas from broadly defined feeding areas; (iii) individuals in a population converge on the mouth of their home stream with remarkable timing; (iv) the rate at which migrating salmon move between sites at which they are marked to sites at which they are recaptured indicates that they are actively orientating. For an enjoyably sceptical view of the precision of orientation of migrating fish, the chapter by Leggett (1984) should be read.

4.11 Implications for exploitation

Many of the commercially most important species are migratory fishes. In a list of the 25 marine species that gave the highest yield to world fisheries in 1978, 24 were migratory species (Harden Jones, 1981). Before the arrival of the Europeans, the Pacific salmon were a staple diet for the Amerinds of the north-west Pacific. Subsequently, the same group of species became the object of major fisheries and the associated canning industry. The Atlantic herring and cod are both migratory species and both have supported major industrialized fisheries in Europe (Cushing, 1988).

A characteristic of many migratory species is that breeding takes place on spatially restricted breeding grounds, examples include the headwaters, used by salmonids and the gravel banks used by Atlantic herring, yet the weight of fish that is generated from a given spawning is far greater than could be accommodated on or close to those breeding grounds. This amplification depends of the fish moving, in some cases long distances, to other richer and larger feeding grounds. It is probable that a crucial factor in the evolution of the migratory habit is the increased growth rate that the migrant fish achieve on the feeding grounds compared with that of non-migratory fish. This increased growth rate will allow them to move rapidly through the sizes most vulnerable to predation (chapter 3). It will allow them to take larger and more profitable food items (chapter 5). Larger females are more fecund (chapter 6) and increased size may give males an advantage in inter-male competition for mates (chapter 6).

However, the migratory habit may make them vulnerable to intensive fishing. In the case of the salmon, the returning fish converge on localized coastal areas adjacent to their home streams. Many migrating fish move in shoals that are easy to detect by commercial fishing boats using electronic methods and once detected the shoal presents a high density of fish to the fishing gear. The migration may also take the fish through fishing areas regulated by different countries or economic groupings, which will make management of the fishery more difficult because there are more interests to satisfy (chapter 7).

CHAPTER FIVE
FEEDING AND GROWTH

5.1 Introduction

From its food a fish must obtain both macro- and micronutrients (Halver, 1989). The macronutrients are proteins, lipids and to a lesser extent carbohydrates. These macronutrients supply the basic building blocks—amino-acids from proteins, fatty acids from lipids and sugars from carbohydrates—which are used to repair damage to body tissues and to synthesize new flesh. The macronutrients are also the fuel that is oxidized during respiration, yielding energy. This energy is used to do the work of maintaining a functioning body, swimming and synthesizing new tissue in the form of body growth or reproductive products (eggs or sperm). The micronutrients are the essential vitamins and minerals that are required, in small quantities, for effective metabolism.

The need for micro- and micronutrients can, potentially, be met from many sources. In teleosts, the wide-range of potential foods is reflected by a wide range of diets and methods of feeding. The agnathans and chondrichthyians show much less variety in their feeding ecology.

The range of potential foods that are available reflects, to a large extent, the physical environment. The range taken by a species depends on its ability to detect, aquire and process the food. Primary production by photosynthesizing green plants is the initial source of virtually all the macronutrients and organic micronutrients eventually eaten by fishes. However, in contrast to terrestrial environments in which herbivorous vertebrates are frequently a conspicuous part of the fauna, herbivorous fishes are often not major components of aquatic food webs. One reason for this lies in the basic properties of aquatic environments. Most of the primary production within the aquatic environment is generated by single-celled algae, maintained in the lit zones of the water column by the buoyancy of the water. These algae are small, typically 2–200 μm in size. A fine mesh or filter is required to harvest them. Once harvested, the cells must be broken open. Typically, fishes exploit these algae indirectly by feeding carnivorously on herbivorous invertebrates or on other invertebrate and vertebrate carnivores. Anchored plants, both algae and flowering plants, are restricted to narrow fringes in shallow water at the edges of rivers, lakes and seas. Even here, an unstable substrate scoured by water currents or tidal surges will often prevent rooted plants from becoming established. Herbivores form significant components of fish assemblages where there are hard substrates that are colonized by encrusting algae, or in shallow waters where macrovegetation can persist, or in highly fertile waters where the phytoplankton reaches high concentrations.

Food webs based on decaying material with its associated flora of bacteria, fungi and other micro-organisms also include fishes. While some fishes can feed on decaying matter and organic detritus, most fishes feed either directly or indirectly on detritivorous invertebrates. The original source of the decaying matter may be within the community, its source is autochthonous, or it may be material that originated in another community but has found its way into the aquatic community. Such material is called allochthonous. Thus, the decaying leaves, stems and roots of water lilies in the littoral of a lake form an autochthonous source of detritus but the leaves that fall into the lake from fringing trees are an allochthonous source of nutrients and energy, as are any terrestrial arthropods or other animals that fall into the water.

A useful term in discussing the feeding ecology of fishes is guild (Begon et al., 1989). Those fishes in a habitat that feed at the same trophic level, exploiting the same class of resources in a similar way, constitute a guild.

5.2 Feeding ecology in riverine environments

5.2.1 *Feeding ecology in the rhithron*

In headwaters, the often torrential flow prevents the development of rooted vegetation and the development of abundant phytoplankton. Primary production is generated by algae and mosses that encrust stones and rocks in the stream bed. As the riffle-pool structure develops, there is an increase in the diversity of the feeding habits of stream fishes.

This progression is illustrated by a stream in eastern Kentucky, USA (Lotrich, 1973). A single cyprinid, *Semotilus atromaculatus*, lives in the topmost stretches. It feeds on terrestrial invertebrates that drop into the stream, an allochthonous material, rather than on food generated within the stream. As the stream order increases, the number of fish species also increases. The first additions feed on aquatic insects, and one species, the stone-roller, *Campostoma anomalum*, eats algae growing in the stream. These are both autochthonous sources of food. With further expansion of the stream, two more feeding guilds appear. A group of three species, all centrarchids, feed on aquatic vertebrates, while the cyprinid, *Pimephales notatus*, has detritus in its stomach. The biomass and abundance of the feeding guilds gives a first approximation of their relative importance in the stream community (Table 5.1). A similar pattern is seen in a warm water stream in Illinois (Schlosser, 1982). The shallow, unstable habitats are occupied by generalized insectivores while, in the deeper and more stable pool habitats, the dominant feeding guilds are the insectivore-piscivores and the benthic-insectivores.

A pattern of an increasing importance of autochthonous food downstream is also shown by the fish assemblage of a small tropical river, the Sungai Gombak in Malayasia (Bishop, 1973). In the small headwater tributaries, *Betta pugnax* (Anabantidae) feeds on terrestrial insects and spiders and on aquatic insects. Two other species present are omnivorous. Further downstream, other feeding guilds appear, including carnivores feeding on aquatic invertebrates and vertebrates. Although the contribution of autochthonous foods increases, the dominant energy supply is from allochthonous plant material and, to a lesser extent, terrestrial invertebrates.

The amount of shading by overhanging vegetation has an important effect on the feeding ecology of stream fishes. Where streams are heavily

Table 5.1 Estimates of abundance and biomass of feeding guilds of fishes defined by their typical diets in 1st-, 2nd-, and 3rd-order stream in eastern Kentucky. Data from Lotrich (1973).

Stream order	Feeding guild	Abundance (per linear m)	Biomass (g dry weight per linear m)
1	Terrestrial invertebrates (1 species)	2.46	1.47
2	Terrestrial invertebrates (1 species)	1.44	0.86
	Aquatic invertebrates (5 species)	1.08	0.26
	Aquatic primary production (1 species)	0.13	0.25
3	Terrestrial invertebrates (3 species)	1.28	1.03
	Aquatic invertebrates (5 species)	1.55	0.63
	Aquatic primary production (1 species)	0.55	0.60
	Aquatic vertebrates (3 species)	0.08	0.72
	Detritus (1 species)	0.05	0.01

shaded, there is little primary production within the stream and the main energy inputs are the leaves, flowers, fruit and other debris dropping from the terrestrial vegetation and the terrestrial invertebrates that get trapped in the surface film. The dominant feeding guilds in such streams are the omnivores that can take both plant and animal material and the carnivores feeding on terrestrial or aquatic invertebrates. In small Panamanian streams, algae grows on the stream bed in unshaded stretches. This algae is eaten by a guild of algivores consisting predominantly of loricariid catfishes (Angermeier and Karr, 1984; Power, 1987). In the Panamanian streams, the biomass of the algal feeders and the species feeding on terrestrial plant material increases with stream size, while that of species feeding on aquatic invertebrates declines.

A feature of most stream fishes is their ability to take many different sorts of food, so they can respond to changes in the availability. This flexibility may be displayed in different ways in different riverine environments. In a Panamanian stream, the diets of the species became less similar in the dry season when food resources diminished and the fish were crowded into restricted areas (Zaret and Rand, 1971). In contrast, in streams in northern South America, the diets became more similar in the dry season as more species switched to ingesting detritus as their preferred foods diminished (Lowe-McConnell, 1987). This flexibility in the feeding of stream fishes is reflected in their lack of highly specialized morphological or physiological adaptations to feeding. The complexity of food webs in tropical freshwater streams is described and analysed quantitatively in Winemiller (1990).

5.2.2 Feeding ecology in flood-plain rivers

The habitat diversity of the flood plain (chapter 2) is reflected in the diversity of feeding guilds. Lowe-McConnell (1987) has classified the feeding guilds in the Zaire River in Central Africa (Table 5.2). The two habitats that contain most guilds and most species of fish are the littoral zone of the river channel and the bays and sloughs. Similar guilds are also described for the River Niger in west Africa (Welcomme, 1986).

When such rivers flood, the inundated areas provide potentially rich feeding grounds. Drowned terrestrial invertebrates become available to the carnivores and omnivores, the nutrients from the soil may support blooms of phytoplankton and the associated zooplankton. When the flooded area is forest, the trees yield another harvest (Figure 5.1). Many fishes of the Amazon basin move onto the forest floor and feed on fruits,

Table 5.2 Number of genera in feeding guilds in habitats of Zaire River (Africa). Data from Lowe-McConnell (1987)

Guild	Open waters	Marginal waters	Swamps	Streams
Mud feeders	2	1	1	
Detritivores	4	3	4	1
Omnivores	4	14	2	7
Herbivores:				
Algae		2		
Macrophyte		6		
Planktivore	2	2		
Carnivores:				
Allochthonous material	2	4	2	5
Benthic insectivores	4	10	5	7
River margin carnivores		9		4
Mixed carnivores	3	9	1	2
Piscivores	3	2		1
Fin biters	1	2		1

seeds, flowers and leaves (Goulding, 1980; Goulding *et al.*, 1988). The species include large characids like *Colossoma macropomum*. For such species, the period of forest flooding is the main feeding season and during the dry season, when the river is confined to the main channel, their stomachs may be largely empty with the fish relying on fat reserves laid down during the flood.

Detritivorous fishes can account for a high percentage of the biomass of the fish fauna of some tropical rivers (Bowen, 1984). In the great rivers of South America, the characid families of Prochilodontidae and Curimatidae are abundant and use the detritus on the flooded floor of the forest and in floodplain pools.

Flooded areas, bays and backwaters are important nursery feeding areas for the young of riverine fishes. The sluggish waters allow the development of a rich flora and fauna of microscopic organisms suitable as food for larval and juvenile fishes (Mills and Mann, 1985).

The feeding guilds of fishes in temperate rivers are less diverse than in tropical rivers. The fauna tends to be dominated by omnivores like the roach, generalist carnivores like the perch and piscivores like the pike (Figure 2.6). Although it is convenient to classify flood plain fishes into feeding guilds, many of them retain the flexibility of diet that is a characteristic of the fauna of headwaters and streams.

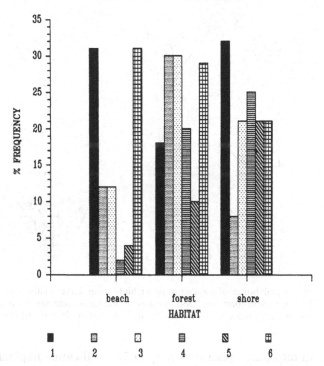

Figure 5.1 Feeding guilds in three common habitats of lower Rio Negro, a flood plain river, in South America. Habitats are river beach, flooded forest and woody shore. Frequencies are the percentages of species feeding significantly on each category. Key to food categories: (1) fish; (2) terrestrial invertebrates; (3) aquatic invertebrates; (4) terrestrial plant material; (5) aquatic plant material; (6) detritus. Data from Goulding *et al.* (1988).

5.3   Feeding ecology in lakes

The importance of allochthonous material of terrestrial origin varies with lake area, shape and the complexity of the shoreline. In small lakes, such inputs may provide an important base for the food webs (Hanlon, 1981) but, in large lakes, autochthonous primary production will predominate.

The Great Lakes of Africa, Malawi, Tanganyika and Victoria are characterized by the high species diversity of their cichlid assemblages. This diversity is associated with a wide adaptive radiation in feeding ecology. Witte (1984) recognized about eleven trophic categories in the haplochrome cichlids collected in the Mwanza Gulf of Lake Victoria. This rich fauna has subsequently been severely depleted by the Nile perch (*Lates*

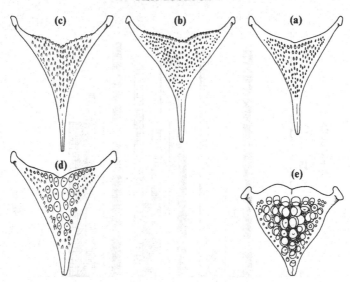

Figure 5.2 Pharyngeal bones of haplochrome cichlids from Lake Victoria showing the relationship between tooth form and food: (a) detritivore; (b) insectivore; (c) piscivore; (d) molluscivore/insectivore; (e) molluscivore. Redrawn from Greenwood (1984).

niloticus), an introduced piscivore (chapter 7). Snail-eating haplochromes, the molluscivores, could be divided into three sub-categories by their mode of preying on snails. Some species grabbed the soft part of the snail before it was retracted into the shell and extracted the snail. Some crushed the shell with their oral teeth. Others crushed the shell between the pharyngeal bones in the throat (Figure 5.2). Two sub-categories of algivores were identified. One group scraped algae from plants (epiphytic scrapers), the other group from stones (epilithic scrapers). The carnivores included insectivores, zooplanktivores, prawn eaters, crab eaters and cleaners that picked ectoparasites off other fishes (chapter 3). The piscivorous haplochromes were divided into three sub-categories. One group fed on whole fishes, some usually on non-cichlids, others on cichlids. A second group fed on the eggs and fry of other cichlids. These paedophages (child-eaters) had various techniques for causing mouth-brooding cichlids (chapter 6) to disgorge their young, which could then be eaten. A third group fed on the scales scraped from the flanks of other fishes. Other categories were detritivores, phytoplanktivores and macrophytivores. In trawl samples in the Gulf, the catch was dominated by insectivores, zooplanktivores and detritus and phytoplankton feeders. Piscivores made

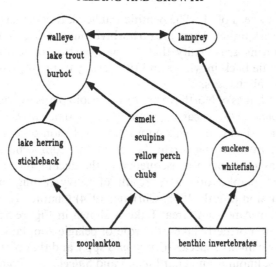

Figure 5.3 Simplified food web for Lake Superior. Redrawn from Cohen and Stone (1987).

up less than 1% of the catch, although this trophic category contributed the highest number of species. Although in many cases, the position and shape of the mouth and the dentition of a species could be correlated with its usual food, the cichlids still retained a trophic flexibility, switching to other foods if they became abundant.

In Lake Tanganiyka, there are endemic herring-like species (clupeoids) exploiting the pelagic zone feeding on both phyto- and zooplankton when young and zooplankton when adult. One species, *Limnothrissa miodon*, has a more generalized diet when adult, eating insects, prawns and young clupeids (Lowe-McConnell, 1987).

In temperate lakes, similar trophic categories can be recognized, but the flexibity in feeding already noted for many tropical species is even more pronounced. This tends to blur the distinctions between the categories (Keast and Webb, 1966; Keast, 1978). In Lake Opinicon, in Ontario, the pelagic zooplankton is eaten by the brook silverside, *Labidesthes*, but this also takes insects from the surface. Two species of sunfish, the bluegill and the pumpkinseed, are generalist carnivores feeding in the littoral. The larger pumpkinseed can feed on snails, crushing their shells with its pharyngeal bones, whereas the bluegill feeds on insect larvae zooplankton. The carnivorous rock bass, *Ambloplites rupestris*, and yellow perch take an increasing proportion of fish as they grow in size. Two piscivores, the northern pike and the largemouth bass, become piscivorous in the first few weeks of life. The benthic bullhead, *Ictalurus*, feeds at night on benthic invertebrates. There are no herbivorous species, which probably reflects the strong seasonality that is characteristic of temperate lakes. Macrophytes die back in winter and the density of phytoplankton shows marked seasonal changes.

In the Great Lakes of North America, a clupeoid planktivore, the alewife, *Alosa pseudoharengus*, became extremely abundant in the 1960s and dominated the fish catches. This population explosion of a planktivorous fish that had invaded the Great Lakes in the 20th century from the east coast, was probably a consequence of the decline of the indigenous piscivores and planktivores as a result of heavy fishing pressure and lamprey predation (Smith, 1968; Christie, 1974) (chapter 7). A simplified food web for an American Great Lake is shown in Figure 5.3.

In even more northerly lakes, the role of pelagic zooplanktivore tends to be taken by whitefish species (*Coregonus* spp.), and that of the generalist carnivore by salmonids and char (*Salmo* and *Salvelinus*) (Svardson, 1976). No species occupy the molluscivore, herbivore or detritivore trophic categories, but each of these food categories may be ingested by a range of species as part of a more generalized diet.

5.4 Feeding ecology in estuaries

For those animals that can tolerate the demanding physical conditions, estuaries offer highly productive feeding grounds. A major energy source is detritus that is imported into the estuary or derived from the decay of estuarine plants, such as salt marsh plants, sea grasses and mangroves. The fishes feed predominantly on invertebrates that are in or closely associated with the substratum, both the meiofauna and the macrofauna. The meiofauna, that is invertebrates with an adult body size less than 45 μg dry weight, is used by smaller fish, typically less than 60 mm in length (Gee, 1989). The most frequent prey are haracticoid copepods (Crustacea) whose activity may make them more vulnerable to the fish than other components of the meiofauna. As the fish grow they use components of the macrofauna such as polychaete worms, larger crustaceans, molluscs and small fish. Juvenile flatfish will crop the siphons of bivalves and the tails of lugworms (Trevallion *et al.*, 1970).

Detritus (and its associated flora and fauna) is also used by some fishes in estuaries and coastal salt marshes. Species vary in the effectiveness with which they can use this abundant but poor-quality food source. Mullet including *Mugil cephalus* feed on detritus (Odum, 1970), although juvenile mullet may be less well-adapted to using this resource than the adults (White *et al.*, 1986). The mummichog, although it remains in salt marshes throughout the year and ingests detritus at times of food shortage, lost weight and had a high mortality on a detrital diet (White *et al.*, 1986).

5.5 Ecology of feeding of marine fishes

5.5.1 *Feeding of rocky- and coral-reef fishes*

The structural complexity of these shallow-water, marine habitats has permitted a wide range of ways of making a living for fishes and this is reflected in the trophic ecology of reef fishes. Most notably, herbivores can form an important component of the fish fauna.

A range of feeding guilds can be identified in fish assemblages on coral reefs (Sale, 1980; Longhurst and Pauly, 1987). Some species feed on or close to the surface of the reef. Others use the coral for shelter but feed in the water column or in habitats away from the reef including adjacent beds of sea grass. The food web of the reef community includes as major inputs primary production by phytoplankton in the water column, primary production by encrusting algae and the algal symbionts of the coral and detritus derived from both plant and animal sources. The encrusting algae is exploited by several herbivorous groups, including grazers such as damselfishes (Pomacentridae) and butterflyfishes (Chaetodontidae), which can bite pieces from the algae and browsers such as blennies (Blennidae) and parrotfishes (Scaridae), which scrape the algae from the substrate.

The zooplankton that grazes the phytoplankton is eaten by clupeoids and damselfish in the water column. The coral is exploited by browsing or grazing fishes including butterflyfishes and parrotfishes. Some detritus is eaten by mullet but detritivorous invertebrates are eaten by a variety of carnivorous fishes, including butterflyfish, grunts (Pomadasyidae) and snappers (Lutjanidae). At the top of the food web are the piscivores. These include species that are adapted for living on the reef and characteristically are either deep-bodied like the groupers or elongated like the cornetfish (Fistularidae). Modes of attack include lying in ambush for passing prey and stalking prey taking advantage of the cover provided by the reef or by other fish. The cornetfish has even been described as forcing its elongated, abrasive snout down the throat of moray eels and feeding on the eel's stomach contents or gut (Johannes, 1981). In the water column

Table 5.3 Percentage of species in seven feeding guilds on coral reefs in the Pacific and Caribbean. (Data from Longhurst and Pauly, 1987)

Guilds	Sites		
	Pacific (Marshall Islands)	Pacific (Hawaii)	Caribbean (Virgin Islands and Puerto Rico)
Detritivores	2	1	0
Herbivores	15	5	13
Planktivores	4	17	13
Omnivores	7	8	10
Coral feeders	4	6	0
Benthic carnivores	56	52	42
Mid-water carnivores	12	11	22
Number of species	225	107	212

adjacent to the reef live pelagic piscivores including sharks, jacks and barracudas, which will attack reef fish when they become vulnerable.

For the One Tree Reef on the Great Barrier Reef in Australia, estimates were made of the proportion of the total weight (biomass) of fish on the reef that were represented by different feeding groups (Goldman and Talbot, 1976). Planktivores represented 10% and grazers including both coral and algal eaters a further 18% of the biomass. Carnivores accounted for the rest, with piscivores making up 54% of the total. The relative proportions of species in seven feeding guilds on reefs in the Pacific and Caribbean are shown in Table 5.3.

Rocky reefs in temperate zones are less diverse. The guild of coral feeders is absent and herbivorous species do not extend much beyond the 40° latitude. A northern example of a herbivore is the Mediterranean blenny, *Blennius sanguinolentus*. Omnivores are exemplified by the Californian cottid, *Clinocottus globiceps*, which feeds on algae, sea anenomes and other invertebrates. Carnivorous reef fishes feed on sessile invertebrates like barnacles and mussels and on mobile invertebrates including gastropods, crabs, prawns and other crustaceans (Gibson, 1969, 1982; Grossman, 1986).

5.5.2 *Feeding ecology on the continental shelf*

The proximity to land and the shallowness of waters over the continental shelves ensures that they are productive. The resource bases for the fishes are provided by the primary production of the phytoplankton, organic

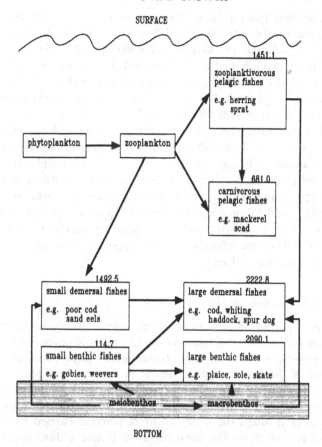

Figure 5.4 Simplified food web for North Sea including estimates of biomass of feeding guilds (× 10³ tonnes). Recalculated from Yang (1982).

material in the form of sinking phytoplankton, the bodies of dead animals and the waste products of the living, and organic material of terrestrial origin. Where cold, nutrient-rich waters upwell, as along the coast of Peru, there is particularly high primary production.

The pelagic zone is usually dominated by clupeoid fishes and their predators. The clupeoids feed either on phytoplankton directly or on zooplankton or on both. The clupeoids and other small pelagic fish and large invertebrates are the prey of the pelagic carnivores including sharks, jacks and mackerel. Demersal fishes take prey that are swimming just above the bottom—the benthopelagic prey—or fish and larger invertebrates that live on the substratum, or they take prey that live in the substratum—the infauna. Small fish, typically less than 60–80 mm long, take meiofauna, especially harpacticoid copepods. Larger fish, including large juvenile and adult rays and flatfish, eat the macrofauna including molluscs and polychaetes. A simplified food web for the North Sea assemblage is given in Figure 5.4, which includes crude estimates of the biomass present in each feeding guild (Yang, 1982).

On the continental shelf of the Gulf of Guinea (West Africa) and including both reefs and soft deposits, Longhurst and Pauly (1987) recognized several feeding guilds: herbivores (7 species); zooplanktivores (11 species); reef epifauna browsers (13 species); benthic infauna carnivores (18 species); carnivores of active benthos (6 species); carnivores taking active benthos and fish in coastal waters (12 species); carnivores of active benthos and fish in offshore waters (17 species); omnivores taking fish, crustaceans and benthic infauna (29 species); pelagic piscivores (34 species) and benthic piscivores (9 species).

5.5.3 Feeding ecology in the open oceans

Open ocean permits fewer ways of making a living because of the homogeneity of the environment and the low rates of primary production. The major energy input is primary production by phytoplankton in the lit upper zone of the water column. Below the zone where photosynthesis can occur, the energy source is provided by organic material imported from above or laterally from the productive coastal zone. Some of this material sinks through the water column, some is carried by turbidity currents that flow down the continental shelf and a third source is the active migration of animals between the more-productive upper waters and the less-productive deeper waters.

In the epipelagic zone, most fishes are generalized carnivores taking their smaller prey by chase and grab techniques. The zooplanktivores include some species that spend their life within that zone such as the flying fishes and some species that migrate at night-time from the deeper mesopelagic zone to feed, returning to the deeper waters during the day (chapter 4). Epipelagic piscivores, including the tunas and dolphin fishes, feed predominantly on the residents rather than the migrants. Large sharks are the top carnivores including in their diet tunas, smaller sharks and cetaceans (Figure 5.5) (Longhurst and Pauly, 1987).

In the mesopelagic zone, three feeding guilds are recognized (Marshall, 1979). The small-jawed carnivores feed on zooplankton and include the

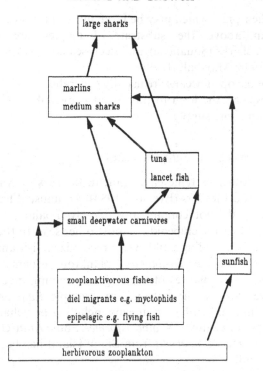

Figure 5.5 Simplified food web for tropical, epipelagic system. Redrawn from Longhurst and Pauly (1987).

species that migrate vertically to the epipelagic zone to feed; examples include small gonostomatids and lantern fishes (Myctophidae). Larger predators with swim-bladders include *Astronethes*. The third guild consists of carnivores that lack a swim-bladder; which includes some stomiatoids.

At the even-deeper and more-impoverished level of the bathypelagic zone, the predators show reductions in body muscle and skeleton, which can be interpreted as adaptations to an energy-poor environment. Characteristic bathypelagic carnivores are the ceratioid angler fish with their bioluminescent lures (Marshall, 1979).

In contrast, the fish associated with the bottom, the benthopelagic and benthic species have well-developed skeletal and muscle systems suggesting that, even under the open ocean, the interface between water and sediment is, at least in comparison with the waters above it, rich in resources. Some species are benthic carnivores taking the benthic in- and epifauna. Such species include chimaeras, halosaurs and eel-pouts (Zoarcidae). Other species take their prey, which may be large invertebrates or fish, from the water column above the substrate. Examples are the rat-tails (Macrouridae), sharks (Squalidae), and deep-sea cod (Moridae) (Sedberry and Musick, 1978; Marshall, 1979).

The fishes of the open oceans are usually generalist rather than specialist feeders. This reflects the need to take what is available in environments where food is in short supply.

5.5.4 *Feeding ecology of Antarctic fishes*

The fishes of another extreme environment, the seas of Antarctica, also tend to be generalist feeders (Everson, 1984). An unusual feature of these seas is that they do not contain any densely-shoaling, planktivorous, pelagic fishes such as the clupeoids of most other seas. In the seas around South Georgia, four feeding guilds were recognized: fish and krill eaters, for example *Dissostichus eleginoides*; plankton feeders, for example, *Notothenia larseni*; carnivores of the larger-swimming invertebrates and fishes that live close to the bottom, for example *Raja geogianus*; and carnivores feeding on benthic organisms including polychaetes, molluscs and crustaceans, for example *N. gibberiformes*. Some Antarctic species will also eat macroalgae. *N. neglecta* from Signy Island has been found with seaweed in its gut.

5.6 Detection and selection of food

Most of the research on the detection and selection of food has been on visually-hunting carnivores, with less attention paid to other senses or other diets.

5.6.1 *Detection and selection in detritivores and herbivores*

With a food that is not mobile, the problem of detection is probably not acute. The major problem confronting the fish is selection. The detritivorous cichlid, *Sarotherondon mossambicus*, feeds selectively on high-quality detritus (Bowen, 1984). The cues used by the fish to make the selection are unknown. Herbivorous fishes are also selective (Horn, 1989). The diets of coral-reef herbivores grazing or browsing on algae do not reflect the proportions of algae available. These herbivores feed in the day so visual selection of plants that differ in colour and shape may play a role. Algae may also be rejected because their secondary compounds make them unpalatable. The texture of the algae can also be relevant with fishes avoiding tough plants of low digestability.

5.6.2 *Detection and selection in carnivores*

Many carnivorous fish detect their prey by sight and the visual characteristics of the prey also play a role in determining whether it is eaten or rejected. Experimental studies suggest that the important visual characteristics of prey are contrast with the background, size, movement, shape and, in some cases, hue.

The maximum distance at which a fish can detect a potential prey is the reaction distance. This depends on the size of the prey (the distance is greater for larger prey), the turbidity of the water and the resolving power of the eye of the fish. A fish swimming through the water column can be thought of as visually searching through a cylinder, whose dimensions are determined by the reaction distance and the rate at which the fish swims (Eggers, 1977). Larval fish can search only small volumes because of their slow swimming speeds and the poor resolving power of their eyes. Larval clupeoids may search less than a litre per hour (Blaxter and Hunter, 1982). If more than one prey are present, the one that appears largest to the fish is the one likely to be attacked (O'Brien *et al.*, 1976). The apparent size of a prey depends on its absolute size and how close it is to the fish.

Studies using different prey types help to define the cues that fish use to select prey visually. The threespine stickleback shows a preference for prey that are red, fast-moving, long and narrow, and larger (where large is defined in relation to the mouth size of the predator and the size range of the prey available) (Ibrahim and Huntingford, 1989). These preferences are not absolute and the stickleback is a generalist feeder whose diet is largely restricted by its own small size (less than about 80 mm in length).

At night, or in turbid water or for prey at distances greater than the visual field of the fish other sensory information will become important. Much less is known about non-visual feeding. Some fish might detect their prey aurally. Sharks may be attracted to the sound made by a thrashing swimmer because it resembles the sound made by a large, but damaged fish (Budker, 1971). Movements of a prey will also set up low-frequency pressure waves that are detected by the lateral line system. Blinded deep-water sculpins, *Cottus bairdi*, fed effectively on moving prey, but this response was inhibited if the lateral line system was blocked (Hoekstra and Janssen, 1985). Some fish can detect the electrical fields generated by living prey. Sharks and rays using their electroreceptors, the ampullary organs, can find prey including flatfishes buried in the substrate (Bleckmann, 1986). A few teleosts—particularly Gymnarchidae, Gymnotidae and Mormyridae—can detect prey through the distorsions created in the weak electrical field generated by the fish (Bleckmann, 1986). (The electric eel of South America, in addition, can stun its prey by a high-voltage discharge from its electric organ.) Many species that feed on benthic prey have chemoreceptors on appendages such as barbels or fin rays and probably use them to detect prey at or just under the surface of the substratum. In the gurnards (Triglidae), the bottom rays of the pectoral fins are free from the fin and are used as feelers in search of benthic food.

Extraordinary feats of prey capture are achieved by a group of gymnotids collected from the main channel of the lower Orinoco River in Venezuela (Lundberg et al., 1987). These fish feed on zooplankton. They somehow detect and capture prey less than a millimetre long in fast-flowing turbid, discoloured water. Neither vision nor electrolocation of the prey are likely given the lack of light penetration and the small size of the prey.

5.6.3   Optimal foraging

Food items might be selected in relation to their value to the fish. A body of theory, optimal foraging theory, has been developed to predict the diet composition of animals on the assumption that the foraging animal maximizes its gain of some currency (Stephens and Krebs, 1986). In other words, food is chosen to maximize the profitability to the forager. The currency in which the profit is measured is usually assumed to be the net rate of energy gain, This is the total energy gained per unit time by the forager less the energy costs of foraging. This will clearly depend on the energy content of each prey type, the time it takes to encounter a prey of a given type and the energy costs of searching for and capturing that prey type. In its classic form, the theory makes simple explicit predictions (Hart, 1986). When encountering prey sequentially and randomly, the forager should select prey in descending order of profitability until including a prey of lesser profitability causes a decline in the overall net energy gain. The diet selection depends only on the density of the more profitable items and not on the density of items that it is not profitable to include in the diet.

Qualitatively, fish can forage in a way that approximates the predictions of the theory; however, quantitatively, the fish do not achieve the predicted, optimum diet composition. The fish continue to include less-profitable prey items in their diet, even after the density of more-profitable prey has reached a level that the former should be ignored. Bluegill sunfish feeding on different size categories of *Daphnia* included more of the larger and more profitable *Daphnia* in their diet as the density of the larger prey increased but also continued to eat some small, less profitable *Daphnia* (Werner and Hall, 1974). Nevertheless, the use of optimal foraging theory allowed Werner and his co-workers to predict successfully seasonal changes in the size -spectrum of prey taken by bluegills in a lake (Werner and Mittelbach, 1981). The simple form of the optimal foraging theory has to be modified to take into account the effects of changes in the motivation of the forager, the risks the forager runs as it forages and its behavioural capacities (Stephens and Krebs, 1986).

5.7 Ecomorphology of feeding

5.7.1 *Filter feeders*

The ammocoete larva of the lamprey burrows in soft sediments in fresh waters. It feeds on algae such as desmids and diatoms and on detritus. The larva maintains a unidirectional flow of water over the filtering apparatus by alternate contractions and expansions of the branchial basket. The water passes from the mouth into the pharynx through a network of branched cirrhi, which exclude any large particles. Smaller particles are then trapped in mucus secreted by the gills and pharyngeal wall. The flow of water is slow and a high proportion of the particles can be trapped (Hardisty and Potter, 1971).

In the filter-feeding clupeoids and other teleosts, the filtering usually takes place on projections from the gill arches, the gill rakers, which are fine and closely spaced. In the gizzard shad, *Dorosoma cepedianum*, the effectiveness with which particles of different sizes are filtered from the water is related to the inter-raker distance (Mummert and Drenner, 1986). As the fish grow, the distance tends to increase, so larger fish are less effective at filtering the smaller particles. As the shad grow, phytoplankton becomes less important in their diet. The basking and whale sharks also filter out zooplankton on their gill rakers. Those of the basking shark are probably shed at the beginning of winter when the shark moves into deeper water (Stevens, 1987).

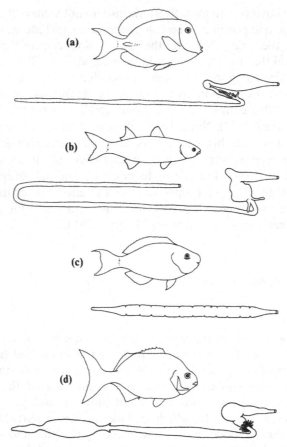

Figure 5.6 Gut adaptations of herbivorous fishes: (a) thin-walled stomach generating low pH, *e.g.* surgeonfishes; (b) muscular, grinding stomach, long intestine, *e.g.* mullet; (c) well-developed phayngeal bones, no stomach, *e.g.* parrotfishes; (d) hind gut with caecum containing microflora, *e.g.* Kyphosidae. Redrawn from Horn (1989).

5.7.2 *Detritivores*

Gill rakers may also play a role in sorting food in detritivores. Mullet sort the sediment in their oral cavity and discard the inorganic material while retaining the organic detritus on the rakers (Bowen, 1984). Some detritivores have muscular grinding stomachs. In the detritivorous cichlids, the stomach is thin-walled, but capable of generating extremely low pH values in the gastric juices. The intestine in detritivores usually has a surface area greater than in either herbivores or carnivores. This is achieved either by a long intestine—in the cyprinid *Labeo* the intestine is 15–21 times the body length—or by having well-developed mucosal folds as in *Prochilodus*.

5.7.3 *Herbivores*

Herbivores share with detritivores the problem that their food is difficult to digest. To obtain useful nutrients, herbivorous fish must break open the walls of the plant cells. Four solutions to the problem can be recognized (Figure 5.6) (Horn, 1989). In groups that include herbivorous cichlids and certain surgeonfishes (Acanthuridae), the stomach is thin-walled and produces highly acidic gastric juices. The intestine is long. The shape of the jaw teeth allows the fish to crop the plant material. In other surgeonfishes and mullet, the stomach is thick-walled and acts as gizzard in which the material is ground up. In parrotfishes, the teeth are fused to form a parrot-like beak. There is a strongly developed mill in which the food is ground between pharyngeal bones as it passes through the pharynx and, although there is no stomach, the intestine is moderately long. The fourth type is uncommon, but is found in the southern hemispheric Kyphosidae which feed on brown seaweeds. In these fish, there is a long intestine and, at their posterior end, there is a hind-gut caecum. This has a well-vascularized wall which contains a community of micro-organisms. It is assumed that the micro-organisms help to break down the plant material and produce materials that can be absorbed by the fish. The solution is analogous to the mechanism used by ruminant mammals.

The long gut allows the fish to process large quantities of food. In some species, the food passes through the gut quickly and the fish obtains its nutrients and energy by processing large quantities of poor-quality food rapidly (Pandian and Vivekanandan, 1985). This would not be true of species that rely on symbiotic micro-organisms to help them digest the recalcitrant plant material.

5.7.4 *Zooplanktivores*

When the zooplankters are small and at high density, some zooplanktivorous fish filter feed. They swim though the aggregation of prey with the mouth open and the prey are collected on the fine, closely-spaced gill rakers. In laboratory experiments, northern anchovy fed like this on brine shrimp (*Artemia*) nauplia (Blaxter and Hunter, 1982). For larger prey, the fish feeds by striking at individual particles. The mouth of zooplanktivores is typically small and tube-shaped. In some species, the jaws are protrusible and the fish picks up the prey as if it were collecting them with a pipette. The gut of zooplanktivores is typically shorter than in herbivores or detritivores. If pharyngeal bones are present, their teeth are fine and closely spaced (Figure 5.2).

5.7.5 Molluscivores

Although the flesh of snails is of good quality, it has to be extracted from the protective shell. Some fish achieve this behaviourally, by grabbing the exposed foot and head, or the siphons before the molluscs can retract them to safety. Other species have morphological specializations that allow them to crush the shells. In the bottom-dwelling skates, jaw teeth are flattened, forming a crushing pavement. In many molluscivorous cichlids, the shells are crushed between the pharyngeal bones, which bear low, rounded crushing teeth (Figure 5.2) (Fryer and Isles, 1972). The molluscivorous pumpkinseed sunfish differs from the zooplanktivorous bluegill sunfish by having pharyngeal teeth that are broad and blunt rather than fine and densely packed (Keast, 1978).

5.7.6 Carnivores

Most fishes are carnivorous and display a wide range of solutions to the problem of capturing live and often highly mobile prey. Only a few will be mentioned here.

The adult lamprey lacks jaws but is still a formidable predator (Hardisty and Potter, 1971). The lamprey attaches to its prey, typically a bony fish, with its oral disc which acts as a sucker (Figure 1.5). Once attached, the lamprey uses the teeth on the tongue to rasp into the flesh. The body fluids and broken tissue of the prey are then pumped into the pharynx. The suction generated by the disc and tongue prevents the prey from dislodging the attached lamprey.

The cookiecutter shark (*Isitius brasiliensis*) has lips that can act as a suction disc. The shark attaches itself to a larger fish then cuts out a plug of flesh with its scalpel-like teeth (Stevens, 1987). The bottom-living nurse shark, *Ginglymostoma*, uses suction to extract prey from holes. The shark makes a seal with its thick lips, then expands its pharyngeal cavity, creating the suction pressure.

Suction is the basic feeding mechanism of teleosts. By suddenly opening the mouth after the buccal cavity has expanded, the fish sucks water in, and with it the prey. This mode of feeding is seen in its most-developed form in large-mouthed predators like sculpins and angler fish.

Carnivores that take a range of prey tend to have generalist morphological traits. Jaw and pharyngeal teeth, when present, are simple cones; there is also a muscular stomach and a relatively short intestine. At the junction of the stomach and intestine there may be a series of blind-ending sacs, the pyloric caeca, but these are also present in some herbivorous and detritivorous fishes. The salmon and trout (Salmonidae) and the perches (Percidae) are good examples of such generalist carnivores.

Many piscivores show a convergence on a morphology that can be described as pike-like (Figure 2.10). The body is elongated and approximately the same depth along its length. The snout is long and somewhat flatted, with jaws equipped with numerous, sharp, backwardly-pointed teeth. In addition to the pike and its relatives (Esocidae), other fishes that show these traits are some piscivorous cichlids and the barracudas (Sphynaenidae).

At the apex of aquatic carnivory are the large sharks such as the Great White (*Carcharodon carcharias*). The jaws of this and similar species are equipped with formidable triangular teeth with a serrated edge. The teeth of elasmobranchs are derived from placoid scales and are continuously replaced throughout life. The gape is large and the stomach capacious so that large prey can be cut up, swallowed and digested. The largest sharks prey on other piscivores including tuna, medium-sized sharks, and marine mammals including seals and porpoises.

5.5.7 *The ecomorphology hypothesis*

The hypothesis assumes that morphology is closely related to and, so predictive of, mode of life. The relationship between body form and habitat has been described in chapter 2. According to the hypothesis, diet should be predictable from the morphology of the fish, particularly from morphological traits related to feeding such as mouth size, jaw shape and dentition. The hypothesis also predicts that species with similar morphologies have similar diets.

This hypothesis has been studied elegantly by Motta (1988) who compared the morphology and diets of ten species of reef-dwelling butterflyfishes. The species included a zooplanktivore, browsers on corals, a browsing omnivore and a benthic omnivore. The range in mouth and

FISH ECOLOGY

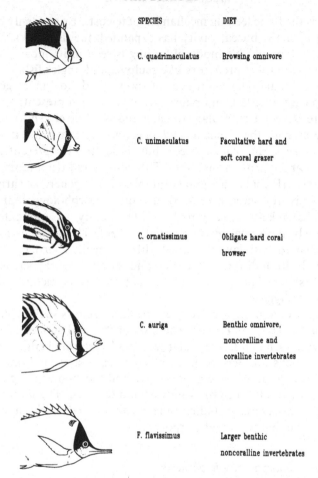

SPECIES	DIET
C. quadrimaculatus	Browsing omnivore
C. unimaculatus	Facultative hard and soft coral grazer
C. ornatissimus	Obligate hard coral browser
C. auriga	Benthic omnivore, noncoralline and coralline invertebrates
F. flavissimus	Larger benthic noncoralline invertebrates

Figure 5.7 Relationship between size and position of mouth in butterflyfishes in relation to usual diet. Redrawn from Motta (1989).

jaw morphology was striking (Figure 5.7). Motta concludes that morphology is not necessarily a good predictor of diet but is a good predictor of how a fish feeds. The mechanics of feeding are an important determinant of diet. Even species that were morphologically specialized would switch from feeding on their usual foods to take zooplankton when it was abundant.

5.8 Trophic categories of fishes

Assigning fishes to well-defined trophic categories is often difficult because of several characteristics of their feeding behaviour. Many species show major changes in their diet during their lifetime—ontogenetic changes. These changes can usually be related to the increase in size with age; this allows the fish to take larger prey. Ontogenetic changes can take place abruptly. The piscivorous pike and largemouth bass switch early in life from a diet of invertebrates to one of fish, a preference that they then retain for the rest of their life. In contrast, perch show a slower change in diet, becoming progressively more piscivorous as they increase in size. In some populations, the perch may never reach a size at which fish become an important component of the diet. In other species, the ontogenetic change relates to the mode of food gathering. In some clupeoids, including the Peruvian anchoveta, the young are predominantly zooplanktivorous. As the fish grow, the spacing between the gill rakers lessens and the adult anchoveta includes a significant portion of phytoplankton in its diet.

Many species are generalist feeders so the diet changes with seasonal changes in food availability, or as the fish moves into different habitats or even because of a sudden super-abundance of a suitable food item. Understanding and predicting these changes in the trophic behaviour of fishes will be one of the keys in developing a comprehensive understanding of the role of fishes in aquatic communities.

5.9 Utilization of food

The principles of the utilization of food are common to all fishes, irrespective of where and how they live.

5.9.1 The energy budget

The energy in the food is either used in the process of metabolism to do useful work, including tissue repair and maintenance, food processing and swimming, or it is stored as the chemical energy of accumulated tissue. The laws of thermodynamics mean that the input of energy must be equal to the output and that as energy is used to do useful work some will be dissipated as heat (Brafield and Llywellyn, 1982). This can be expressed in a simple equation:

$$A = R + P$$

where A is the energy of the assimilated food, R is the heat lost in the process of metabolism and P is the energy gained by the fish in the form of growth. The energy in the assimilated food, A, is less than the energy in the food consumed, C, because some of that energy is lost in the evacuated faeces (F) and some is lost in the nitrogenous excretory products produced in the metabolism of protein (U). Consequently, the following equations describe the energy budget:

$$A = C - (F + U)$$

$$A = R + P$$

where (A/C) is a measure of the efficiency with which consumed food is assimilated. In terms of weight of food consumed, the efficiency is greater for carnivores than for herbivores or detritivores because the latter ingest relatively large quantities of indigestible material. (P/C) is a measure of the efficiency with which consumed food is converted to growth, it is gross growth efficiency. Growth efficiency can also be expressed in terms of assimilated food, (P/A).

Compilations of the energy budgets of carnivorous and herbivorous fishes allowed Brett and Groves (1979) to draw up average energy budgets, which give a crude picture of how the energy in the food is partitioned. For carnivores, the average was:

$$100C = 44R + 29P + 27E$$

and for herbivores:

$$100C = 37R + 20P + 43E$$

where $E = (F + U)$.

These figures suggest that it is slightly more costly energetically to be a carnivore (44R *versus* 37R) but that carnivores assimilate their food more efficiently (27E *versus* 43E).

5.9.2 *Nutrient budgets*

Food provides nutrients as well as energy and although the constituents of foods can be broken down and new compounds synthesized or eliminated, the budgets of nitrogen, phosphorus and other elements must be described by the balanced equation:

$$income = retained + eliminated$$

The retained portion is the amount that is incorporated into the body of the fish. The eliminated portion is lost to the fish, but excretory products of the fish may be important to other animals and plants in the community as a source of nutrients. Coral-reef fishes that feed off the reef in patches of sea grass or mangrove swamps import nutrients like nitrogen and phosphorus into the reef system, releasing them in faeces and soluble excretory products such as ammonia.

For a group of species living in lowland streams in Poland, Penczak (1985) estimated that for every kilogram of element retained, the fish had to consume about 7.9 kg of carbon, 6.6 kg of nitrogen and 3.1 kg of phosphorus. As a percentage of the dry weight of the fish, carbon accounted for about 42%, nitrogen for about 10% and phosphorus just over 2%. These nutrients are locked up in the biomass of the fish.

5.10 Rate of food consumption

As the energy and elemental budgets make clear, the energy and material that can be used in activities, R, and growth, P, will depend on the rate of income in the form of food. The rate of food consumption will depend on the availability of food in the environment and on the readiness of the fish to feed.

The relationship between the number of food items taken by a fish and their abundance in the environment is described by the functional response curve. This is typically asymptotic (Figure 5.8). There is a density of food beyond which there is no further increase in the rate of feeding, the fish is feeding at its maximum rate.

This maximum rate is, in turn, related to the size of the fish and the environmental temperature. As a fish grows bigger its absolute requirement for food increases but, in many species, the food consumption per unit weight declines. In relative terms, as fish get bigger they tend to consume less. This can be described by a simple model, where C_{max} is the maximum rate of consumption and W is fish weight:

$$C_{max} = nW^m$$

and m usually takes a value of less than 1.0, indicating that the relationship is allometric (Figure 5.9).

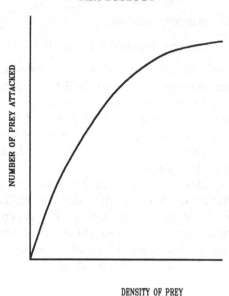

DENSITY OF PREY

Figure 5.8 Idealized functional response curve showing asymptotic relationship between prey density and number attacked.

Over the range of temperature that a fish can tolerate (chapter 2), the rate of consumption first increases up to a maximum but then any further increase in temperature causes a decline in consumption (Figure 2.2b). Other factors such as salinity and pH may also affect the rate of consumption but their effects are less well studied. In an experimental study on brown trout, estimates for the daily maximum consumption at 10°C ranged from 5.3% of body weight for a fish weighing 5 g, to 2% of body weight for a 300 g fish (Elliott, 1975). The equivalent range at 15°C was 11.3% to 4.3%. Observations on a variety of natural populations suggest that a rate of consumption of 0.5–5% of body weight is typical for many species.

5.11 Growth

Most fishes, unlike birds and mammals, continue to grow in length and weight after they have reached sexual maturity, although the rate of growing declines as the fish get bigger. The growth pattern observed is the result of an interaction between a potential for growth defined by the

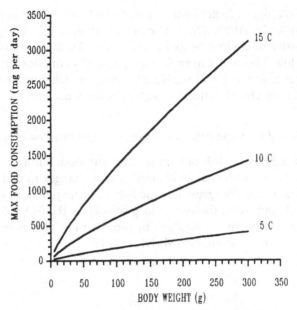

Figure 5.9 Relationship between body weight and maximum daily food consumption for brown trout (*Salmo trutta*) at three temperatures—5°C, 10°C and 15°C. (C = temperature in °C). Curves calculated from data in Elliott (1975).

genotype of the fish and the environmental conditions experienced by the fish (Figure 1.7a). Food availability and temperature are often the most important environmental identities. Weatherley and Gill (1987) provide a comprehensive review of the biology of fish growth.

5.11.1 Body sizes of fishes in relation to habitat

Adult size in fishes covers a wide range. The smallest is about 10 mm in a tropical goby (Miller, 1979) while the largest is the whale shark whose length can reach 15 m or more. These differences are caused by differences in growth patterns, which even within a species can vary considerably.

In the total fish fauna of an area, the proportion of fishes with a typical mature size of less than 100 mm in length decreases from the tropics to the northern polar zone in both fresh waters, shallow marine waters and deep (> 2000 m) marine waters (Lindsey, 1966). The importance of small fishes in the mesopelagic zones, especially small myctophorids and stomiatoids, is described by Marshall (1979). However, Eversen (1984) noted that many of the endemic species of Antarctica were also relatively small. The ecological factors that favour small body size in fishes are discussed by Miller (1979). One obvious advantage is the ability to reach sexual maturity in environments in which the absolute amount of food available is low. The ability to exploit a structurally complex environment such as a coral or rocky reef is also an advantage (chapter 2). However, the causes of the global patterns observed remain unclear.

5.11.2   *Effect of food and other environmental identities on growth*

Growth depends on the fish obtaining sufficient food, but as a fish grows it can often take a wider range of prey. These changes in diet can then have effects on the further growth of the fish. In some generalist carnivores such as perch and trout, the switch to piscivory as the fish grows in size is often followed by an acceleration in growth rate. This is because the larger prey are more profitable (Mittelbach, 1983).

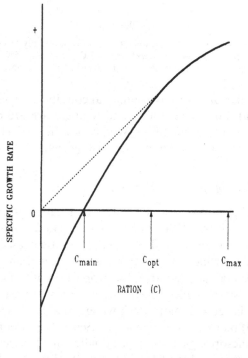

Figure 5.10   Schematic diagram of relationship between ration and specific growth rate. Dotted line indicates tangent from origin identifying the optimum ration. C_{main} = maintenance ration; C_{opt} = optimum ratio and C_{max} = maximum ration.

The relationship between growth rate and consumption rate in fish is usually a negatively accelerated curve (Figure 5.10). A minimum quantity of food, C_{main}, is required just to maintain a constant body size. A consequence of the shape of the relationship between growth rate and food consumption is that the maximum growth efficiency is achieved at a ration, C_{opt}, that is less than the maximum rate of consumption, C_{max}. The fish cannot simultaneously maximize its growth rate and its growth efficiency (Brett, 1979).

The growth rate also depends on the profitability of the food (Elliott, 1979). There is a size range of food particles that allow the highest growth rate with this rate declining if the food particles are smaller or larger than this optimum range (Wankowski and Thorpe, 1979). Growth rate also varies with temperature. In the optimum temperature range, fish that are fed unrestricted rations have their highest growth rate (Elliott, 1979) (Figure 2.2b). For brown trout, a temperate stenotherm, the optimum temperature for growth is 12–13°C.

Growth is also affected by other abiotic factors including oxygen and salinity and by biotic factors (Brett, 1979). The abiotic and biotic factors that provide the best growing conditions vary with and reflect the mode of life of the species. The presence of other fishes can reduce growth rates through interference or exploitation competition (chapter 3). Where fish are crowded together on fish farms, the fish often have to be graded for size at regular intervals to separate the larger dominant fish from smaller subordinates. In some shoaling fishes, the presence of conspecifics may enhance growth rates.

5.11.3 *Relationship between weight and length*

So far growth has been discussed largely in terms of weight (or energy content), but fish grow in length as well as in bulk (Weatherley and Gill, 1987). The relationship between weight, W, and length, L, for fish from a given population is usually well-described by the equation:

$$W = aL^b$$

where b is 3.0 if the fish is growing isometrically (without changing shape). If the fish changes shape as it grows (allometric growth) b will differ from 3.0 (Figure 5.11). If the fish gets relatively thinner, b is less than 3.0, if it gets plumper, b is greater than 3.0. The equation relating weight to length thus gives some indication of the condition of a fish in a population. One frequently used, if crude, measure of condition, CF, is:

$$CF = W/L^3.$$

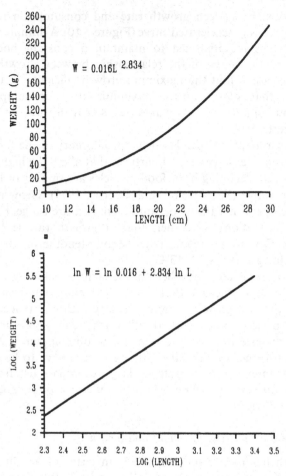

Figure 5.11   Relationship between weight and length for brown trout (*Salmo trutta*). (Upper graph, values plotted arithmetically; lower graph, values plotted logarithmically.) Curves based on data in Elliott (1975).

Fish in good condition will have higher values of CF than those in poor condition.

5.11.4   *Measurement of growth*

Growth can be measured as the change in the absolute weight (or energy content) or length of the fish over time. If the size of a fish is measured at time intervals that are sufficiently long to obscure shorter-term changes in size caused by seasonal or reproductive cycles, the usual growth pattern is an asymmetrical S-shape. In fisheries biology, the pattern of growth over the lifetime of a fish is frequently described by the von Bertallanfy curve, which takes the form:

$$L_t = L_\infty (1 - e^{(-K(t-t_0))})$$

where L_t is the length of the fish at age t, L_∞ is the asymptotic length of the fish, t_0 is the hypothetical time at which the length of the fish is zero, K is the rate at which the growth curve approaches the asymptote. Although originally derived from physiological principles, the equation is best regarded as providing a useful, mathematical summary of the growth curve (Beverton and Holt, 1957; Ricker, 1979). An example is shown in Figure 5.12. The von Bertallanfy curve is less suitable for describing the early growth of fish or when the time intervals between successive measurements of growth are small.

The absolute growth rate is a function of fish size. A fish that increases in weight by 1 g from 1000 g to 1001 g has the same absolute growth rate

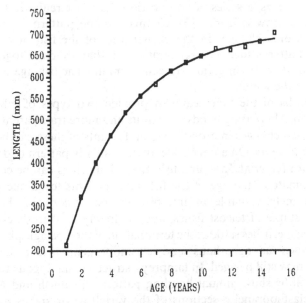

Figure 5.12 Growth in length of walleye (*Stizostedion vitreum*) from Lake Ontario. A von Bertallanfy growth curve has been fitted to the data. Curve parameters: $L_\infty = 712$ mm; $K = 0.0232$; $t_0 = -0.566$ years.

as a fish increasing in weight from 1 to 2 g over the same time period, yet the latter has doubled its weight. A measure of growth rate that is independent of size is the specific growth rate, which can be measured by:

$$(\log_e S_f - \log_e S_0)/t$$

where S_f and S_0 are the final and initial sizes of the fish over a defined time period and t is the length of the time period. During the lifetime of a fish, the specific growth rate is usually high during the early life-history stages and decreases as the fish gets older and larger. If a fish is not growing, its specific growth rate is zero.

5.11.5 *Ageing of fish*

In the laboratory the growth of a fish can be measured over a known time period. However, in natural populations, the growth patterns usually have to be determined from the sizes of fishes of known age. Fortunately, many fish ages can be ascertained because of changes in the pattern of deposition of calcium salts in structures such as scales, the otoliths of the inner ear, fin spines, vertebrae or other bones (Summerfelt and Hall, 1987).

In teleost scales, a series of concentric ridges or circuli are laid down as the scales grow (Figure 5.13). Changes in the pattern of the circuli indicate seasonal changes in the growth rate of the fish or events like spawning. Patterns such as a series of circuli that are close together and which recur at yearly intervals are called annuli. The fish age is assessed by counting the annuli.

The otoliths of the inner ear can provide two types of information (Figure 5.13). Alternating bands of opaque and more transparent material mark seasonal changes in growth rate and counts of these can be used to age the fish in years. On a finer scale, there is a daily pattern of deposition of the otolith material. In young fish, these daily rings can be counted to give an estimate of the age of the fish in days. This technique promises to be extremely valuable in interpreting the ecology of the earliest life-history stages of teleost fishes, and is already used to determine the age at which reef fishes settle at the termination of their pelagic phase of life.

Fishes such as agnathans and elasmobranchs, which lack bony structures, may still be aged. In lampreys, structures analogous to otoliths called statoliths show annual banding patterns (Beamish and Medland, 1988). In elasmobranchs, sections of the vertebrae or spines may show annuli. A different technique for ageing fishes is to use the frequency distribution of lengths. If lengths are measured for a large sample from a

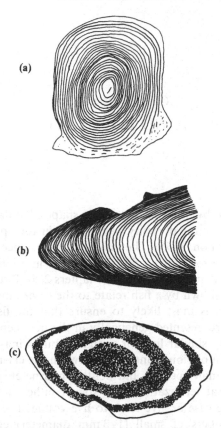

Figure 5.13 Structures used to age teleost fishes: (a) scale of salmonid fish showing circuli; (b) otolith of young cichlid showing daily rings; (c) otolith of flatfish showing annual rings.

population that has a seasonal breeding pattern, each age class present in the population should show up as a peak (or mode) in the length-frequency distribution. An age class will be separated from the preceding younger and the succeeding age class by the equivalent of growth in length between successive breeding seasons (Pauly and Morgan, 1987).

Care must be taken to validate whichever method of ageing is used routinely for a population. This can be done by comparing the results of ageing by several techniques and by laboratory studies on the factors that control the formation of annual or daily patterns.

CHAPTER SIX

LIFE-HISTORIES AND POPULATION DYNAMICS

6.1 Introduction

The diversity shown by fishes in their body shapes, habitats and ways of feeding is matched by the diversity of their life-history patterns (Breder and Rosen, 1966; Balon, 1975). The adaptive significance of body shape and traits related to feeding is being interpreted in relation to where a fish lives and how it makes its living (chapters 2, 5). But how does the life-history pattern shown by a fish relate to the environment in which it lives? What pattern is most likely to ensure that the fish leaves some offspring and so is represented genetically in the next generation?

Adult fishes living similar lives in similar habitats may have different life-history traits. An obvious contrast is between the elasmobranchs and teleosts who share many pelagic and benthic marine environments. The former have internal fertilization and spawn either a few large eggs protected by an egg case or give birth to live young. The latter typically produce large numbers of small (1–3 mm diameter) eggs, which are fertilized externally. Consequently, the mortality rates between fertilization and sexual maturity are far higher for teleosts than for elasmobranchs. In North Sea plaice, 80% of eggs and larvae die per month, with over 99% of eggs failing to reach the late larval stage. The life-history of elasmobranchs is in some ways closer to that of mammals or birds than to teleosts.

Even closely related species living in similar habitats can show difference in their life-history patterns. In almost all the Pacific salmon of the genus *Oncorhynchus*, the adults die after breeding just once—the semelparous condition. Salmon and trout belonging to the closely related genus *Salmo* are iteroparous. A portion of the adults survive to breed again. These examples show that fishes living in similar habitats and exploiting similar food resources have evolved different ways of ensuring that copies of genes carried by individuals in one generation are present in the succeeding generation.

Life-history patterns depend not only on the ecological circumstances prevailing at the time and in the recent past but also on the history of the gene pool, which is represented by a population at that time. Present life-histories must be interpreted in the context of both ecological (contemporary) and phylogenetic (historical) constraints. The long, separate evolutionary histories of the elasmobranchs and bony fishes are reflected in their different modes of reproduction.

A further consequence of differences in life-history patterns between species is likely to be differences in how populations respond to environmental change by changes in their numerical abundance and biomass. An understanding of the processes that cause changes in the abundances of populations is particularly important for those who have to manage populations being exploited by fisheries (chapter 7).

6.2   Life-history traits and the concept of trade-off

6.2.1   Fecundity and egg size

The total volume of eggs that a fish can produce will be limited by the space available in the body cavity to accommodate the eggs (or developing embryos) just prior to spawning. The total egg volume will be determined by the number of ripe eggs (fecundity) and the volume of an individual egg. In fishes there is no general tendency for larger fish to produce larger eggs, so fecundity increases with body size (Bagenal, 1978; Wootton, 1979).

Table 6.1 Examples of fecundity, egg diameter and fish length taken from freshwater fauna of the UK

Species	Adult Length (mm)	Fecundity (no. of ripe eggs produced)	Diameter (mm)
Gasterosteus aculeatus	40–80	80–400	1.4–1.6
Rhodeus amarus	50–60	40–100	3.0
Cottus gobio	50–100	50–700	2.0–2.5
Gobio gobio	80–140	1000–3000	1.5
Gymnocephalus cernua	120–150	1000–6000	1.0
Leuciscus leuciscus	150–200	3000–27000	2.0
Rutilus rutilus	100–250	5000–100000	1.0
Perca fluviatilis	150–250	4000–100000	2.0–2.5
Tinca tinca	250–500	100000–600000	0.8–1.0
Salmo trutta	200–1000	200–10000	4.0–5.0
Esox lucius	400–1000	50000–100000	2.5–3.0
Salmo salar	500–1200	5000–20000	5.0–7.0

This relationship holds both within species, and other things being equal, between species (Table 6.1). Within a species, fecundity (F) increases approximately with the cube of body length (L), that is:

$$F = aL^b$$

where a and b are parameters and b is approximately equal to 3.0 (Figure 6.1). Since total egg volume is determined by fecundity and the

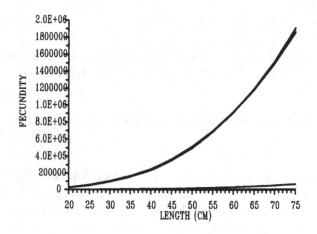

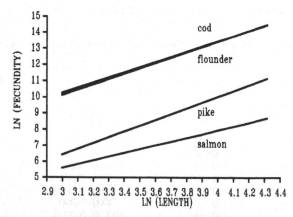

Figure 6.1   Relationship between fecundity and length for four species, cod (*Gadus morhua*), American plaice (*Hippoglossoides platessoides*), both marine spawners, pike (*Esox lucius*) and salmon (*Salmo salar*), both freshwater spawners. (Upper graph, values plotted arithmetically; lower graph, values plotted logarithmically.)

mean volume of an egg, there tends to be a trade-off between fecundity and egg size (Wootton, 1984b; Elgar, 1990). This trade-off between number and size has already been noted in the comparison between teleosts and elasmobranchs. However, even within the teleosts, the trade-off is observed. When sexually mature, cod and salmon may be of similar size (1 m in length) but the salmon will produce a few thousand large eggs (diameter 4–7 mm), whereas the cod will spawn several hundred thousand smaller eggs (diameter 1.5 mm). A simple calculation will demonstrate that these represent similar total egg volumes.

Larger eggs give rise to bigger post-hatch larvae. Mouth size, swimming capacity and sensory capabilities increase with size and so fish that hatch at a larger size are able to detect and exploit a larger range of prey items and are less susceptible to predation. For evolution to favour larger eggs, the gains in survival that accrue through being hatched at a large size must compensate for the reduction in number of progeny spawned.

6.2.2 Serial or total spawning

Some fishes, called serial or batch spawners, increase the number of eggs that they produce in a breeding season to levels much higher than would be predicted from their body size by spawning eggs in a series of clutches. A female threespine stickleback is physiologically capable of spawning more than ten clutches in a breeding season, although each clutch at the time of spawning will represent about 20% of her total weight. The northern anchovy may spawn twenty times or more over a year. In contrast many species, for example pike and perch, are total spawners releasing all the eggs that will be produced in a breeding season over a few hours or days. The next batch of eggs will not ripen for at least a year. A likely cost of serial spawning is that the growth in size of the fish will be slowed or stopped during the period that eggs are being produced. A second cost may be increased mortality (Pitcher and Hart, 1982).

6.2.3 Age and size at maturity

Ecological theory predicts that, other things being equal, reaching sexual maturity at an early age is a potent way of increasing genetic representation in the succeeding generation (Begon et al., 1989). However, it is at the cost of becoming mature at a small size and consequently having a low fecundity and also probably at risk from a wider range of predators than fish that mature at a larger size. The size range at which fishes become mature ranges from about 10 mm for a miniature goby to over 2000 mm for species such as the long-lived sturgeons and sharks. In fishes, there is a general tendency for fish that have a high growth rate to become mature at an early age, have a low asymptotic size and a high mortality rate (Pauly, 1980).

Within species, the mean age and size at which maturity is reached varies between populations and environmental conditions (Alm, 1959). A common but not universal pattern is for fishes with a high growth rate to have a reduced age at maturity. This reduction may or may not be accompanied by a reduction in the size at maturity.

6.2.4 Post-spawning survival

In semelparous species, no adults survive to breed a second time. An iteroparous life-history means that some individuals survive to breed again and again. In iteroparous species, the chances of surviving to breed again can vary from being less than 10% in most populations of Atlantic salmon, to being high, as in the North Sea plaice, which, between the ages of 5 and 15 years, has a survival rate of about 90% per year. A disadvantage of iteroparity is that fewer resources are invested in each spawning attempt than by semelparous species because the adults of iteroparous species must invest some resources to ensure their own survival. A comparison of semelparous and iteroparous anadromous salmonids and shads suggested that the iteroparous forms used less than 60% of their body reserves when making a spawning run whereas the semelparous forms expended 70% or more (Glebe and Leggett, 1981).

6.2.5 Reproductive life-span

Iteroparous species that reach sexual maturity at a large size and consequently have a delayed age-at-maturity usually have a longer reproductive life-span than species that reach maturity at an earlier age. This relationship has been observed in both the clupeids and the flatfishes (Roff, 1981). A consequence is that in long-lived species an individual will breed at several ages and, in any one year, fish of many ages will be breeding in a population.

6.2.6 Parental care

Elasmobranchs have internal fertilization and this is often associated with viviparity (Wourms et al., 1988). In the viviparous forms, which represent about 65% of living elasmobranchs, there are a variety of ways in which the developing embryo obtains nutrients. Probably the most primitive condition is when the embryo obtains all its nutrition from the egg yolk while in its mother's uterus (yolk feeding or lecithotrophy). During development, the embryo loses mass as the yolk is metabolized. Consequently, the viviparity confers protection against predation during the egg stage but does not result in larger hatchlings than oviparity. Benthic electric rays, *Torpedo*, and the benthopelagic spiny dogfish, *Squalus acanthias*, show this pattern. The order Lamniformes includes many pelagic sharks. In some of these, the embryo obtains nutrients by cannibalizing other eggs (egg feeding or oöphagy) or embryos (embryo feeding or adelphophagy). By this cannibalism, the embryo can greatly increase its mass before birth. In the mako shark, *Isurus paucus*, only one embryo at a time develops to term in each uterus. At birth, the young mako are nearly 1 m long and weigh over 5 kg. (A typical teleost larva will hatch at a length of about 5 mm.) Embryonic growth also occurs in those species that have a maternal-embryonic connection through which the mother passes nutrients to the embryo. About 27% of viviparous sharks are known to be placental. In these species, the increase in weight of the embryo from fertilization to birth may range from 6000–10 000%. The spadenose shark, *Scoliodon laticaudus*, ovulates eggs less than 1 mm in diameter but at birth the pups are 130–150 mm long. There is no evidence of post-birth, parental care in the elasmobranchs.

In the bony fishes, parental care in the form of viviparity is rare, although it has evolved in several lineages. Of the 20 000 + species of teleosts, some 500 are viviparous. In viviparous teleosts, the ovary itself is the site of gestation, with the embryos developing either in the ovarian lumen or in the ovarian follicle. As in the elasmobranchs, the source of nutrition for the developing embryo ranges from a dependence on egg yolk to a trophic connection with the mother. In the cyprinodonts, the viviparous guppy (*Poecilia reticulata*: Poeciliidae) shows a 25–45% reduction in embryonic mass during gestation, while in the Goodeidae the embryonic mass increases by up to 15 000%.

A cost of viviparity is that, in most cases, the fecundity of the female is much reduced from that which might be expected for an oviparous species of the same size. Again the trade-off is between increased egg and juvenile survivorship and reduced fecundity. An exception is the viviparous rockfishes (Scorpaenidae) (Wourms, 1991). The Scorpaenidae show a range of conditions from a typical teleost pattern of spawning large numbers of small, pelagic eggs to viviparity. This has evolved in one sub-family, which includes the genus *Sebastes*. Fertilization is internal and the eggs develop in the ovarian lumen. However, there is no significant reduction in the fecundity of these viviparous rockfishes. The larvae are born at less than 10 mm in length although with well-developed organ systems. During gestation, the developing embryos may obtain some nutrition from the breakdown of dead eggs and embryos, as well as from their own yolk. The pregnant rockfish has to provide the oxygen required by the many thousands of developing embryos and dispose of their waste products.

In egg-laying (oviparous) species, parental care may be limited to the preparation of a spawning site prior to egg release. Lampreys dig a nest pit by moving stones with their body or carrying them away using the suctorial mouth. After spawning the eggs are covered with stones. A similar pattern is seen in stream-spawning salmonids. Once the eggs are covered no further parental care occurs.

Post-fertilization care of eggs and larvae occurs in 22% of teleost families, although not neccessarily in all the species in a family (Sargent and Gross, 1986). The commonest form is paternal care (Table 6.2). Often the parental care is guarding, with the parent protecting the eggs from predators. In examples such as the sunfishes, the sticklebacks and some wrasse, the male builds a nest in which the eggs are laid and then guarded. Eggs laid on the substrate or in a nest are in danger of depleting the water around them of oxygen or of being buried under sediment. Ventilation of the eggs often accompanies egg guarding, with the parental fish wafting a current of water over the eggs. A few families, for example the Cichlidae, include species in which the parent broods the eggs in its mouth—in cichlids usually the mother. The parent gently churns the eggs to keep them ventilated. Even after hatching, the young cichlids remain close to their mother and will dash back into her mouth if danger threatens. In a few teleosts, the eggs are carried around on the surface of the parent (Balon, 1975). In pipefishes and seahorses (Syngnathiformes), the female deposits the eggs in a groove or brood pouch on the abdomen of the male.

Table 6.2 Distribution of parental care in fishes, expressed as percentage of families. Data from Sargent and Gross (1986)

Mode of care	Non-teleosts	Teleosts
Male parental	6	11
Female parental	66	7
Biparental	0	4
No care	28	78

When it is the male that looks after the eggs, there need be no reduction in female fecundity as a trade-off for parental care. It is probable that parental males pay a cost in terms of increased mortality and a reduced growth rate. In those cases where there is female or biparental care, fecundity is reduced. For example, in the biparental cichlid, *Cichlasoma nigrofasciatum*, the interval between successive spawnings is much longer if the female looks after her eggs and young than when the eggs are removed immediately after fertilization.

6.2.7 *Modes of sexuality*

Most fishes have separate sexes—the gonochoristic condition—with internal fertilization in the elasmobranchs and external fertilization in the aganthans and most, but not all, bony fishes. However, other modes of sexuality occur.

6.2.7.1 *Hermaphroditism.*

The condition in which one individual can function both as a female and as a male is rare in the vertebrates, but a few examples are found among the bony fishes (Warner, 1978). In simultaneous or synchronous hermaphroditism, an individual has functional ovaries and testes at the same time so that self-fertilization is a possibility. Several deep-sea fishes are simultaneous hermaphrodites and also some members of the perciform family, the Serranidae, living in shallow sub-tropical and tropical waters. There is also a small cyprinodont, *Rivulus*, found in coastal swamps in southeastern North America, which is a self-fertilizing hermaphrodite. In sequential hermaphrodites, an individual first has functional gonads of one sex and then at some point in the life-history, the gonads undergo a transformation to become functional as the other sex. In protandrous forms, the male phase with functional testes precedes the female phase with functional ovaries. In protogynous species the female phase precedes the male. In some of the sex-changing species there may be individuals, the primary males or females, that do not change sex. The best-studied of these sex-changing, sequential hermaphrodites are reef-dwelling fishes including wrasses (Labridae), parrotfishes (Scaridae) and gobies (Gobiidae) (Warner, 1978, 1984, 1988) (see p. 148).

6.2.7.2 *Parthenogenesis.*

All-female populations in which only maternal genes are inherited from generation to generation are also rare in vertebrates. Parthenogenesis does occur in two genera of cyprinodonts,

Poecilia and *Poeciliopsis,* living in small springs and streams in arid regions of south-western North America (Vrijebhoek, 1984) (see p. 142). These parthenogenetic forms are curious in that they are sperm parasites of co-existing and closely-related, sexually-reproducing species. For the eggs of the parthenogenetic forms to develop, they have to be activated by sperm from males belonging to the sexually reproducing species. In one mode, paternal genes contribute to the genotype of the offspring but only the maternal genes are inherited from generation to generation—a system called hybridogenesis. In the other system, gynogenesis, the sperm only activates the egg. The paternal genes are not incorporated into the genome of the developing embryos.

6.2.8 *Place and timing of spawning*

Reproductive success is also determined by where in the habitat and when in the year spawning occurs. These traits are best considered in the context of the habitat in which reproduction takes place.

6.2.9 *Life-history theory*

Life-history theory seeks to predict the combination of traits that are likely to evolve in given environmental circumstances (Begon *et al.,* 1989). In its present form, the theory suggests that the relative rates of mortality in the pre-reproductive and reproductive phases of a life-cycle are important. If the pre-reproductive mortality rates are high and unpredictable, delayed reproduction, long adult life-span and iteroparity are favoured. If adult mortality rates are high and unpredictable, early reproduction with a high investment in that early reproduction is favoured.

6.3 Breeding patterns of riverine fishes

6.3.1 *Fishes of the rhithron*

In low-order streams, the high and variable flow rates offer both advantages and disadvantages to spawning fish. The advantages include highly oxygenated water whose flow rate keeps sediments from accumulating over the eggs. A major disadvantage is the risk to the eggs and young fish of being swept downstream during spates and killed.

In cold water, upland streams in North America and Eurasia, the commonest species are salmonids, cottids and a few cyprinids (chapter 2). The salmonids bury their eggs in gravel nests through which there is a good flow of water. While spawning is occurring, the area around the nest is defended vigorously—especially by the male. Many male salmonids develop a secondary sexual characteristic in the form of a kype—a hooked jaw armed with fang-like teeth—which is used in fighting other males. In cottids, the male defends a hole formed between or under stones. The female spawns in the hole and the eggs adhere to the walls. The male ventilates the eggs and defends the hole from intruders. Upstream cyprinids such as the European minnow scatter the eggs in crevices between rocks and stones but do not guard the site. In eastern North America, many of the small cyprinids guard the eggs (Moyle and Herbold, 1987).

The eggs of stream-spawning salmonids are among the largest produced by teleosts and the fecundity is correspondingly low (Table 6.1), a few hundred to a few thousand depending on body size. A comparison of egg diameters of temperate marine and freshwater species showed that the freshwater species (including both riverine and lacustrine species) tended to have larger eggs (Figure 6.2) (Wootton, 1979; Elgar, 1990). Larger larvae hatching from larger eggs may have an advantage in coping with water currents and exploiting bigger prey. In running water, phytoplankton and small zooplankton are usually present in low densities.

The fishes that live permanently in these low-order streams are usually small-bodied (chapter 3). They are often serial spawners. Large-bodied fish exploit the habitat for spawning by migrating upstream. Examples include the anadromous lampreys and salmonids, which are total spawners and in many cases semelparous (chapter 4).

Stream salmonids spawn in autumn or winter, and the eggs buried in the gravel develop slowly. Even after hatching, the young, called alevins remain in the gravel as they exhaust their yolk reserves. The salmonid fry only emerge from the gravel in spring and at, for teleosts, a large body size of about 30 mm. In contrast, the cottids, cyprinids and cobitids (loaches) of these low-order streams spawn in spring or summer. The prevailing higher temperatures mean that it takes only days for the eggs to hatch, and the larvae emerge less than 10 mm in length.

A feature of the low-order streams at lower latitudes is the presence of viviparous cyprinodonts. For example, several species of *Poecilia* live in small streams in central and south America. Experimental and field studies of the guppy in streams on Trinidad have supported the predictions of life-history theory (Reznick *et al.*, 1990). At localities where piscivores take mainly adult guppies, guppies mature at an earlier age and give birth to

FISH ECOLOGY

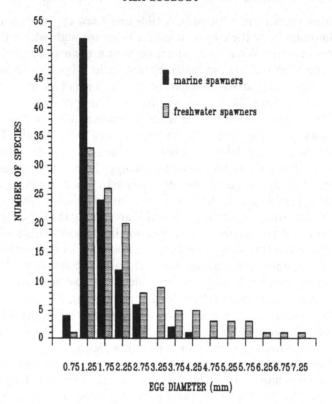

Figure 6.2 Egg diameters for marine and freshwater species from northern temperate waters.

more and smaller young than at localities where the piscivores take mainly juvenile guppies (cf. predictions on page 140).

In the streams of the arid zone of southwestern USA and Mexico live the parthenogenetic forms of *Poecilia* and *Poeciliopsis*. One explanation of the appearance of this unusual mode of reproduction in this environment is the 'frozen-niche' hypothesis (Vrijenhoek *et al.*, 1987). This suggests that parthenogenesis is a mechanism that reduces intraspecific competition in an environment in which there are few other fish species and so low interspecific competition. All-female clones, each one consisting of genetically identical individuals but differing genetically from other clones, are derived from a genetically variable sexual ancestor. Selection between clones produces an assemblage that partitions the food and space in a way that reduces ecological overlap between the clones (chapter 3).

6.3.2 Fishes of the potamon

For many fishes of the potamon, their reproductive biology is geared to the cycle of flooding. Spawning usually takes place at some time between low water and peak flood and only rarely during falling floods (Welcomme, 1985). This timing ensures that the young are hatching into an environment in which the area of feeding habitat is expanding (chapter 5). At higher latitudes, the time of spawning is also correlated with the seasonal increase in temperature and day-length associated with the advance of spring and summer. This period usually corresponds to high water levels as the snows melt.

Total spawners are common among potamon fishes, probably because there is a time-period in the flood cycle that is optimal for the development of the eggs and growth of larvae. In tropical rivers, examples include many of the larger South American characoids and African cyprinids. In more temperate regions, the Chinese carps and other cyprinids, pike and perch are all total spawners. Serial spawners are also present. Often these are smaller fishes with short life-spans. Most potamon species are iteroparous.

The site of spawning can be in the main river channel or in flooded vegetation. In the Madeira River, a tributary of the Amazon, large characins migrate to the main channel to spawn before moving onto the flooded floor of the forest to feed (Goulding, 1980). The young are swept downstream into backwaters and flood-plain lakes. In Chinese rivers, large-bodied cyprinids like the bighead carp, *Aristichthys nobilis*, spawn pelagic eggs in the river; these are carried downstream and into nursery areas along the flooded fringes of the river. The turbidity of the water gives the eggs and larvae some protection from predation, while the movement of water also keeps the eggs oxygenated and free of sediment.

Two other important spawing sites are vegetation and stony substrates (Balon, 1975). Species like the pike, common carp and tench deposit eggs that stick to submerged aquatic plants. Others such as the walleye, *Rhinichthys* spp. and *Catastomus* spp., scatter their eggs on rock and stones. Yet others like the roach, bream and perch, attach their eggs to submerged vegetation, logs or stones.

The risk of predation of eggs and larvae by invertebrates or vertebrates is higher in the potamon than in the rhithron. In total spawners, high numbers of eggs are produced over a short period, which may swamp the predators. Other fishes guard their eggs. In species like the characoid *Serrasalmus*, the guarded eggs are attached to plants. In *Arapaima*, a hollow is made in the flooded ground. Male anabantoids of south-east Asia make a floating nest of bubbles to hold the eggs. This ensures that the eggs are close to the oxygen-rich surface film in still, warm waters prone to deoxygenation. In temperate waters, the male stickleback builds a nest, either on the bottom or in vegetation. The female bitterling, *Rhodeus amarus*, deposits her eggs in the mantle cavity of freshwater mussels through a long, tubular ovipositor.

Most flood-plain species avoid the danger that eggs will be exposed by falling water levels by the timing of their spawning. However, some cyprinodont species of South America and Africa can maintain populations in pools that regularly dry up (Wourms, 1972). These small, short-lived 'annual' fishes lay eggs that can survive in the bottom mud or vegetation as the pool empties. The eggs go into diapause during which egg development is arrested. Further development is then stimulated when the pond refills.

6.4   Breeding patterns of lake fishes

The littoral fishes of lakes share many of the reproductive patterns of flood-plain fishes. Indeed, some lake fishes migrate into the rivers to spawn, taking advantage of environments opened up by flooding (Lowe-McConnell, 1987). In these species and also the littoral species of temperate lakes, spawning is seasonal.

The main spawning period for temperate lake fishes is from spring to mid-summer. There are some notable exceptions to this. The burbot spawns in mid-winter. The lake trout spawns in autumn, while many of the whitefish also spawn in autumn or winter. The eggs are spawned over gravel but, in the case of the burbot and some whitefish, the eggs are slightly buoyant and so are moved gently round by water movement.

In lakes at low latitudes, some species have distinct spawning seasons, whereas other species spawn throughout the year. Even the latter species tend to have periods when a higher proportion of the population is in breeding condition than at other times. The advantages of seasonal spawning in tropical lakes in which there are only small changes in temperature and day-length may, in some species, reflect changes in plankton production, which are linked to wet–dry seasonality or patterns of wind-driven circulation. In lake Jiloa, in Nicaragua, the cichlid species that spawn in the littoral zone show seasonal breeding, which may reflect a response to both food availability and competition for spawning sites. The dominant species, *Cichlasoma citrinellum*, occupies most of the available spawing sites during the wet season. It ceases to breed in the dry season, when other cichlids are able to spawn (McKaye, 1977).

Lacustrine species include both total and serial spawners and, for littoral species, some form of parental care is common. Males of centrarchids like the bluegill sunfish clear a nest site on the bottom in which the female spawns. In the bluegill sunfish, males have a breeding system that includes alternative reproductive strategies (Gross, 1984). Some males mature at the age of 6 or 7 years of age, build a nest and guard the eggs in it. Other males mature as early as 2 years old and at a small size. These males do not attempt to build a nest, but lurk around the nests of large parental males and rush in to fertilize some eggs as the parental male spawns with a female. As they get bigger these 'sneaker' males start to resemble females, which allows them to closely approach the nests of parental males and sneak fertilizations. Alternative mating strategies involving large 'territorial' and small 'sneaking' males are seen in other fishes including some anadromous salmonids (Gross, 1984).

In Africa, many of the lake-dwelling cichlids are mouth-brooders (Fryer and Iles, 1972). The males dig nests on the bottom which, in some species, are elaborate constructions. These nests are clustered together in colonies, with the males defending small territories immediately around the nest. After spawning, the female picks up the fertilized eggs in her mouth and departs to brood the eggs and young fish away from the males.

6.5 Breeding patterns of estuarine fishes

Although estuaries are used by many marine fisheries as nursery areas, only a few species reproduce there. The species that do tend to be small-bodied, serial spawners and some form of parental care is common. Examples include gobies and sticklebacks. A few species take advantage of the tidal cycle to reduce the risk to their eggs of predation by aquatic predators. These species spawn at high tide, depositing their eggs among damp vegetation or in damp sand. As the tide recedes, the eggs are emersed, although still in damp conditions. Hatching occurs at a subsequent high tide where the eggs are once again immersed. Because of the relationship between the tidal cycle and the phase of the moon, these species typically have a lunar spawning cycle. The Atlantic silverside, *Menidia menidia*, spawns around the full or new moon when the tidal range is greatest.

The Atlantic silverside has a relatively unusual sex-determinatioin mechanism (Conover and Heins, 1987). In most vertebrates, the sex of an individual is determined by its genotype. In some *Menidia* populations, sex is determined by an interaction between the sex-determining genes and environmental temperature: an example of environmental sex determination. Most offspring produced at the beginning of the breeding season when water temperatures are low become female. Offspring produced towards the end of the season at high water temperatures become male. This effect is adaptive because the females have a longer growing period and so reach a larger body size than males. In females fecundity increases with body size, whereas in male *Menidia*, there in no evidence that fecundity increases with size. In northern populations in which the breeding season is short and temperature is not a good cue of time, this environmental effect on sex determination is weak or non-existent.

6.6 Breeding patterns in the sea

6.6.1 *Reef fishes*

Many reef fishes are small-bodied, serial spawners. Another common feature is that there is usually a planktonic dispersal phase early in the life-history (Gibson, 1982, 1986; Sale, 1980; Thresher, 1984). Either the eggs themselves are pelagic or the larvae become pelagic after hatching (chapter 3). This planktonic phase has been interpreted as a mechanism to avoid intense predation on the reef, to achieve dispersal of the young or to allow the young to exploit a patchy planktonic food resource. These common features should not obscure the great diversity of reproductive patterns shown by reef fishes (Thresher, 1984).

On temperate reefs, breeding is usually seasonal with most species spawning in spring or summer. A few species, including the cottids of western European shores, spawn in late winter and spring (Fish and Fish, 1989).

On tropical reefs, there is a range from those species that breed only during a restricted time of the year to those that show some spawning throughout the year, although in the latter there are usually peaks in the intensity of spawning (Thresher, 1984; Lowe-McConnell, 1987). Fishes of the coral reefs of Jamaica show a peak of spawning activity when annual sea temperatures are at their minimum (February to April). In contrast, species of pomacentrids on the Atlantic coast of Panama show a minimum spawning output during December to March, which is the dry season (Robertson, 1990).

The seasonal spawning of some reef fishes could be timed to give the pelagic larvae the best chance of returning to the reef given the prevailing winds and currents (Johannes, 1981). However, little correlation was found between the time of maximum spawning and time of maximum recruitment for Panamanian reef fishes (Robertson, 1990).

Coral-reef fishes show five main modes of spawning (Table 6.3), but demersal and pelagic spawners are by far the most common (Thresher, 1984). On average, pelagic spawners are larger than demersal spawners. Consequently, they have higher fecundities. However, if the comparison is made for fish of the same size, there is no significant difference between fecundity nor egg size between demersal and pelagic spawners. Demersal eggs take longer to hatch than pelagic eggs and the larvae from demersal eggs tend to hatch at a more advanced stage of development.

Many reef fishes defend territories, with spawning taking place in the territory of the male. If the eggs are demersal, they are usually laid where the risk of dislodgement by turbulent water is reduced and the eggs are less exposed to predation. Crevices, the underside of stones and holes are favoured places. Some reef fishes are nest-builders, for example some wrasses of temperate waters. Parental care is also common, with the caring parent typically but not invariably being the male, guarding and, in many cases, ventilating the eggs as they develop. The Apogonidae are mouth brooders but it is the male that broods. In pipe-fishes and sea-horses

Table 6.3 Distribution of modes of spawning in reef teleost fishes, expressed as number of families that include species showing particular modes of care. Families may be represented in more than one mode. Data from Thresher (1984)

Mode of care	Number of families	Mean maximum length (mm)	Mean egg volume (mm³)
Pelagic eggs	36	446 ± 5.64	1.45 ± 0.670
Demersal eggs	13	144 ± 3.00	2.17 ± 1.56
Egg scatterers (spawning pelagic, eggs demersal)	2	—	—
Benthic broadcasters (demersal spawners, eggs pelagic)	1	—	—
Live bearers	2	—	—

(Syngnathidae), the female transfers the eggs to a brood pouch or groove on the abdomen of the male where they are fertilized and then develop to or even beyond hatching.

In other reef fishes aggregations form at sites on the reef where groups of males and females congregate for spawning, with the eggs being released into the water column. Some large fishes such as the groupers migrate seaward, forming large spawning aggregations on the seaward side of the reef. The spawning typically takes place on a falling tide and at times of the new and full moon, when the tidal range is greatest. These conditions should flush the eggs away from the reef and so reduce predation on them by reef-dwelling fishes and invertebrates, while also helping to disperse the eggs and larvae (Johannes, 1981; but see also Thresher, 1984).

A species may show both territorial and group spawning depending on the population density and the size distribution of the fish in the population (Warner, 1984). If the population density of mature fishes is high, the rate of intrusion into territories may be so great that they become indefensible and spawning will take place in groups. Small males unable to defend territories will also join in spawning aggregations.

Hermaphroditism, either simultaneous or sequential, is frequently recorded in reef fishes. Sea basses, Serranidae, are simultaneous hermaphrodites, with functional ovarian and testicular tissue present in the same gonad (Fischer, 1986). However, there is no self-fertilization. In some species, pairs form and the partners alternately take on the male and female roles during a spawning sequence. The selective factors that have favoured the evolution of simultaneous hermaphroditism in the Serranids are not understood.

Sequential hermaphroditism is expected to evolve when the sexes differ in the rate at which they gain in reproductive ability with size (Warner, 1984, 1988). If males gain faster than females, a change from female to male is predicted, if the females gain faster than the males protandry is predicted.

In the reef-dwelling fishes that show a protogynous sex-change such as the wrasses, large males are able to sequester a harem of females with which they can mate. Thus reproductive success as a male is dependent on achieving a large body size, and a sex change from being female at smaller sizes to being male at larger sizes is favoured. At high population densities, small males may achieve some reproductive success by taking part in group spawnings. Anemonefishes form monogamous pairs but are protandrous. This probably relates to the effect of an increase in body size on the fecundity of the female. In a monogamous pair, the male will not gain in reproductive ability by an increase in size whereas the female will increase in fecundity. A male anemonefish is also at risk of being displaced by another male. In many sex-changing species, the change of sex is triggered by a change in the social environment, for example, the disappearance of the dominant fish, the male in protogynous species and the female in protandrous species, or by a change in the sex ratio in a group of fishes (Shapiro, 1984).

6.6.2 *Breeding patterns of continental-shelf fishes*

Most shelf teleosts spawn buoyant, pelagic eggs and provide no parental care. Just before spawning, the eggs take up water and, after fertilization, the eggs take up more water, causing them to swell and become more buoyant. Even demersal fishes like the gadoids and flatfishes produce pelagic eggs. The larvae of flatfishes are pelagic before they metamorphose into juveniles with the laterally compressed adult form and adopt a benthic life in on-shore nursery areas.

The obvious exception to pelagic spawning is provided by the shelf elasmobranchs, many of which show parental care in the form of viviparity. The second exception is provided by those teleosts that spawn demersal eggs. These include the herring. In the North Sea, the Atlantic herring spawns on gravel patches, while off the east coast of Canada it spawns, like the Pacific herring, on submerged vegetation. Capelin of the North Atlantic, *Mallotus villosus* (Osmeridae), lay demersal eggs with some populations spawning in gravel in the inter-tidal zone of beaches.

Shelf fishes are typically iteroparous. Some such as the anchovies are serial spawners, while others like the cod, herring and flatfishes are total spawners. Serial spawning is associated with relatively long breeding seasons. However, even those species that have total spawning can have extended breeding seasons, with individual fish spawning at different times in the year. In St. Georges Bay, Nova Scotia, successive cohorts of larval herring are produced throughout summer and autumn, whereas all the mackerel larvae are produced over a 2-month period in June and July with most produced within a 3-week period (Lambert and Ware, 1984). In some species, the timing of breeding is quite precise. In northern populations of cod, plaice and herring, the variation in the date of spawning from year to year is small (Cushing, 1975). Different stocks of herring spawn at different times. Stocks that spawn in autumn or winter produce large eggs but have a low fecundity, whereas spring and late-summer spawning stocks produce more but smaller eggs. This trade-off between egg size and fecundity is related to the feeding conditions in which the newly hatched larvae find themselves. Larger larvae will be at an advantage when the plankton is sparser.

For shelf fishes from lower latitudes, the spawning seasons tend to be longer than for high latitude fishes (Longhurst and Pauly, 1987). Some species, for example the Peruvian anchoveta, show clear peaks in spawning activity, whereas other species show an even pattern of spawning over the year.

The seasonality of spawning probably relates to the requirement of the pelagic larvae for adequate quantities of plankton in a size range that can be ingested. At high latitudes, there is a well-defined annual cycle of plankton production, whereas at lower latitudes the cycle of production is less strongly seasonal. Those species that are serial spawners or which produce a series of cohorts of larvae over the breeding season may reduce the risk of producing all their larvae at a time when planktonic production is poor—a reproductive strategy that has been called 'bet-hedging'.

Many shelf fish are long-lived and so populations of a species contain many age-classes. An iteroparous fish with a potentially long life-span reduces the risk of reproductive failure because the production of young is spread out over several breeding seasons. Poor breeding seasons should be balanced by good ones.

6.6.3 Breeding patterns in the open ocean

The teleosts of the epipelagic zone have similar reproductive patterns to the shelf fishes. They produce large numbers of small, buoyant, pelagic eggs and show no parental care. They are typically iteroparous. The elasmobranchs are usually viviparous, giving birth to a few large offspring at long intervals of time.

Deep-sea fishes seem to fall into three groups (Gordon, 1979; Marshall, 1979). Firstly, there are the deep-sea chondrichthyians that have internal fertilization. Some, such as the chimaeras, lay large eggs protected in a capsule. Others, like the squaloid sharks, are viviparous. A few of the bathybenthic teleosts, including some egg-pouts (Zoarcidae), are also viviparous. Other benthic fish, including the sea-snails (Liparidae) of temperate and polar waters, produce large eggs that probably remain on or near the bottom as they develop. Some of these benthic teleosts may show forms of parental care including mouth brooding. Many benthopelagic, bathypelagic and mesopelagic teleosts spawn bouyant pelagic eggs, which float upwards in the water column. Consequently, the larvae hatch into the more productive waters just below, in or above the thermocline. After a period of growth in these shallower waters, the juveniles move down the water column to the parental habitat in deeper waters.

Marshall (1979, 1984) suggests that the fishes of the impoverished meso- and bathypelagic zones that have reduced skeletal and muscular system when compared with fish from higher in the water column are economizing on investment in the somatic component of the body to maximize investment in gametes. This allows these fishes to produce sufficient eggs to cover the inevitable losses that occur as the eggs float upwards and the larvae develop in productive, shallow waters.

The small fishes that dominate the mesopelagic zone in tropical and sub-tropical waters, including the lantern fishes, may live for only one or two years. At higher latitudes, these small, mesopelagic migrators may live for several years (Childress et al., 1980). The larger fishes of the meso- and bathypelagic zone and the large benthopelagic fishes such as the rat-tails (Macrouridae) are long-lived and are presumably iteroparous. However, Childress et al. (1980) suggest that some bathypelagic fishes only reach sexual maturity late in life and may only breed once (semelparous).

The sparse food supplies in the lower meso-pelagic and bathypelagic zones result in low population densities. This raises the problem of finding a mate. Many mesopelagic fishes are bioluminescent (chapter 3) and have well-developed eyes, thus visual cues probably play a role. Bathypelagic fishes often have reduced eyes, and other sensory modalities such as olfaction may be used. The ceratioid angler fishes show a profound sexual dimorphism. The males reach sexual maturity at a much smaller body size than the females (dwarf males) and are slimmer with well-developed eyes and olfactory organs. The females, as they metamorphose to the adult form, take on a globular shape with a well-developed lure. The dwarf males have large testes and as they become mature develop pincer-like jaws with which they may attach to the female at the time of mating. In some ceratioids, the dwarf male becomes permanently attached to the female living parasitically off her, but also ensuring that she always has access to a mate.

Hermaphroditism is present in some deep-sea fishes, and is probably an adaptation to the problem of finding a mate at low population densities. In the mesopelagic zone, the large-bodied, predatory alepisauroid and notosudid teleosts are synchronous hermaphrodites. The chlorophthalmids, including the tripod-fishes, which are the commonest benthic fishes in deep waters in the sub-tropical and tropical zones, are also synchronous hermaphrodites. The incidence of self-fertilization in these groups is unknown.

6.7 Population characteristics

When considering changes in the abundances of fish populations, their population dynamics, the contrast is less between fishes from different habitats and more between the dynamics of the highly fecund teleosts and the low-fecundity elasmobranchs. The basic principles of population dynamics apply to both groups. A variety of techniques are used to obtain the estimates of population abundance that form the basis for the study of population dynamics. These are described in Pitcher and Hart (1982).

6.7.1 Net reproductive rate

Net reproductive rate (R_0) is a useful index of the success of a population's life-history pattern. It can be defined as the rate of multiplication per generation of the population or, equivalently, as the average number of daughters produced per female born into the population (Begon *et al.*, 1989). (It is assumed that the rate of fertilization is not limited by a shortage of males.) If each female, on average, is replaced by one daughter, the population will be numerically stable, neither increasing nor decreasing. The net reproductive rate is one. If the net reproductive rate is, on average, less than one, the population is declining. If the value is greater than one, the population is increasing.

The value of the net reproductive rate is determined by the prevailing age-specific rates of survival and fecundity in the population. It is calculated as:

$$\text{Net reproductive rate } = R_0 = \sum_{x=\alpha}^{\Omega} l_x m_x$$

where l_x is the probability that a female survives from fertilization to reach age x, and m_x is the fecundity expressed as the number of female eggs produced by a female aged x, α = age at maturity, Ω = maximum possible life-span in the population.

In fishes, both survival rates and fecundity depend on size. Consequently, the growth rate of individuals in a population will affect R_0 through the size-related consequences.

6.7.2 Population production

The abundance, growth rate and survival of individuals will also determine the changes in the biomass of the population over time. Production is a measure of the total amount of tissue synthesized by a population over a defined time interval (Mann and Penczak, 1986). It provides an estimate of rate at which fish flesh is becoming available to exploiters of that population—predators or scavengers. It is defined as:

$$P = g\bar{B}$$

where g is the specific growth rate (chapter 5) and $\bar{B}$ is the mean biomass of the population for the time interval over which production is estimated. The biomass (B) of a population at any given time is calculated as:

$$B = N \cdot \bar{W}$$

N is the abundance of the population and $\bar{W}$ is the mean weight of an individual in the population at that time. Estimates of production for periods that include the spawning season must also estimate the contribution of eggs and sperm.

6.7.3 Density dependence

The pattern of changes in abundance and production by a population will partly reflect the importance of the effect of the population's own density on its mortality rate, fecundity and growth (Begon et al., 1989). If the effect of density is such that at high population densities, the mean R_0 is less than 1 (declining population), while at low population densities the mean rate is greater than 1 (increasing population), the population will, in the absence of other disturbing factors, fluctuate around the density at which the mean R_0 is 1 (Figure 6.3). The potential of a population to show an increase in abundance after its numbers have been reduced is important for fisheries. By harvesting fish from a population, the fishery sets in train density dependent changes in the mortality, birth and growth rates, which compensate for the losses to the fishery (chapter 7).

6.8 Dynamics of fish populations

6.8.1 Stock-recruitment relationship

Although density dependent effects on R_0 tend (usually) to stabilize the numerical abundance of a population, fish populations, to the despair of

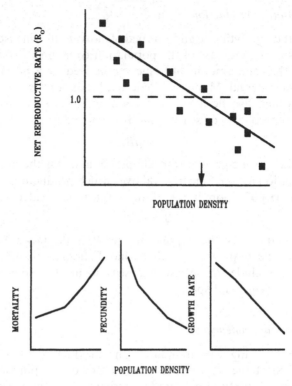

Figure 6.3 Density dependence illustrated by density-dependent relationships for net reproductive rate (R_0), mortality, fecundity and growth. Vertical arrow indicates equilibrium population density at which mean R_0 is equal to 1.0.

fisheries managers, can show wide fluctuations in abundance. This is especially true of teleost fishes. The basic reason is the high fecundity of most teleost fishes. This gives the population the potential for extreme changes in abundance. A long-lived pelagic fish may produce many millions of eggs in a life-time. It requires only small changes in the survival rates between spawning and sexual maturity to generate large changes in the absolute abundance of adults in the population. Consequently, a feature of many teleost populations is the weak relationship between the size of the adult population (the stock) and the number of recruits to that component of the population (Rothschild, 1986).

A knowledge of the size of the adult population does not allow precise predictions of the number of fish that will be recruited to the adult population over time. Much effort has been devoted by fisheries' biologists to describe the stock-recruitment relationship in a way that would be useful for managing fisheries. The studies of Ricker (1954) and Beverton and Holt (1957) have been especially influential.

A second feature of many populations is the presence of dominant-year classes or cohorts. These arise when unusually high numbers of juvenile fishes are produced from a breeding season, often for reasons that are not fully understood. In long-lived fish, the presence of a dominant-year class can often be detected for many years after its first appearance because, at any age, it forms a higher proportion of the total population than would normally be expected for fish of that age. A famous example is the 1904 year class of Atlantic herring, which was still detectable in the population in 1922 (Nikolskii, 1969). The recruitment of the 1963 year class of North Sea plaice was 200% greater than the mean annual recruitment (Rothschild, 1986). In exploited fish populations, the appearance of a dominant-year class will often allow a greater yield to be taken.

The uncertain stock-recruitment relationship and the phenomenon of dominant-year classes highlight the importance of understanding the processes that affect survival between spawning and the onset of sexual maturation. The events that occur in the first few months of life are usually crucial. Predation and food abundance, together with the effects of temperature and water movements, are probably the main factors that affect the survival of young fish (Bailey and Houde, 1989). These factors may interact in complex ways. At high food levels, the fish can grow more quickly and so become less susceptible to predation (chapter 3). If there is sufficient food, growth rates tend to increase with temperature (chapter 5), but higher temperatures also result in higher metabolic rates and hungrier predators.

In some species, the onset of independent feeding after the yolk of the eggs has been absorbed may be a critical period. If the concentration of suitable food for the larval fish is inadequate, the fish will weaken, become more susceptible to predation or will starve. Cushing (1990) has extended this idea in his hypothesis of match–mismatch. This predicts that feeding, growth and survival will be good when there is a close match between the time when the larvae start feeding and the production of food suitable for larval and small juvenile fish. This will lead to good recruitment by that cohort of fish. If the larvae start feeding when food is sparse, recruitment from that cohort will be poor.

The difficulties of sampling larval and juvenile fishes in the sea in a way that provides good quantitative estimates of their mortality and growth rates has made it difficult to test the hypotheses of a critical period and 'match–mismatch', although there is some circumstantial evidence (Rothschild, 1986; Cushing, 1990). The capelin off the eastern coast of Canada shows large year-to-year fluctuations in year-class strength. There is a good correlation between year-class strength and two abiotic environmental variables: the frequency of onshore winds during the period immediately following hatching, and the water temperature experienced during the subsequent larval drift (Leggett et al., 1984). This is one of the few examples where a convincing causal account can be given for the correlation. The onshore winds drive warmer surface water onto the beaches where the capelin spawn. The increase in temperature stimulates the larval capelin to emerge from the gravel—a process aided by the disturbance caused by the turbulent water. This warmer surface water carries a lower density of potential predators of the larvae than the deeper, colder water that upwells when offshore winds blow. The warmer water also has a plankton community which contains more suitable prey for the larvae. In the warmer water with adequate food, the larvae can grow faster (chapter 5) and this will reduce the time they are at risk from some size-selective predators (chapter 3).

The clearest example of a critical period comes from a study of the population biology of trout, Salmo trutta, in streams in the Lake District of England. This study by Elliott (1975, 1979, 1981, 1985, 1989) is an outstanding example of a study of the population dynamics of a teleost. The main analysis was of an anadromous population of trout (Figure 6.4). After hatching, the fish spend 2 years in the stream before migrating to the estuary or sea. The time spent in the sea before returning to spawn varies from under a year to over 2 years. The population is typical of teleosts in that the size of the spawning stock is only a poor predictor of the number of fish that recruit to the spawning stock (Figure 6.5). However, analysis showed that the number of young by the end of the first summer of life was similar, irrespective of the number of eggs spawned. Survival of the eggs buried in the redds was good but, after emergence from the gravel, mortality in the first months was strongly density-dependent. Careful sampling showed that there was a critical period after emergence during which mortality was density-dependent. After this critical period, mortality was not density-dependent. The length of the critical period varied between about 30 and 70 days and was itself density-dependent. The period was shorter the higher the density of the eggs. In this trout population, the main density-dependent regulation of the population was occurring in a critical period after emergence. The growth rate after

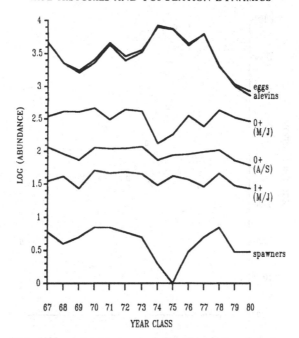

Figure 6.4 Population dynamics of anadromous trout (*Salmo trutta*) population showing abundance for each year class for : (i) eggs; (ii) alevins, (iii) 0 + (May/June) parr; (iv) 0 + (August/Sept) parr; (v) 1 + (May/June) parr and (vi) adult spawners for year classes from 1967–1980. (Year class is year of emergence from gravel not year eggs spawned.) Abundance (per 60 m²) at each stage is given in $\log_{10}$ units. Data from Elliott (1985).

emergence was not density-dependent but during the critical period both large and small trout had poorer survival at high densities than intermediate-sized trout. A key to survival for the trout was the ability to take up a feeding territory in the critical period. At the beginning of the period, there was often a significant number of trout without territories but by the end of the period few.

The characteristic features of the salmonid life-history, including burying the eggs, the large size of juveniles on emergence from the gravel and their subsequent feeding territoriality, make it difficult to generalize the results of this study to teleosts that scatter their eggs and in which the larvae hatch at a small and vulnerable size. This trout study does emphasize the value of the detailed study of the early life-history stages for an understanding of the population dynamics.

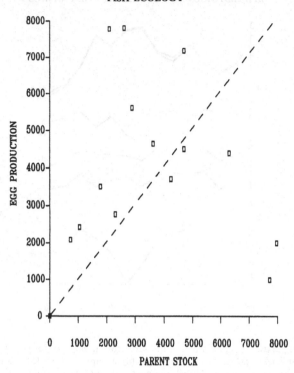

Figure 6.5 Stock–recruitment relationship for anadromous trout population shown in Figure 6.4. Stock expressed as number of eggs present at start of a year class, recruits as number of eggs produced by that year class (eggs per 60 m^2). Dashed line is line of equal replacement, that is, recruits = stock. Note the weak relationship between stock and recruits. Also there is evidence of density dependence—at low densities, recruits exceed stock and at high densities stock exceeds recruits. Data from Elliott (1985).

6.8.2   Long-term changes in population abundance

The Lake District trout population has been studied since 1966. Historical records of fisheries give longer time-series from which the fluctuations in the abundance of fish populations can be inferred (Cushing, 1982). Unfortunately, these data will confound the effects of the fishery, including changes in fishing techniques, fishing effort and the economic value of the catch with the effects of natural abiotic and biotic factors on the abundance of the population. There is some evidence of long-term changes in abundance that may be correlated with large-scale climatic changes. Off the Pacific coast of the USA, the Californian sardine and the northern anchovy seem to have alternated as the dominant pelagic planktivore. Herring populations in north-western Europe have shown periods of great abundance when they have supported thriving fisheries and other periods of scarcity. On a shorter time-scale, Cushing (1990) provides evidence of a possible link between poor recruitment in several demersal populations in the north-west Atlantic and the movement of a large mass of unusually cooler and fresher water across the Atlantic.

An understanding of the effects of climatic conditions on population dynamics will only come from a better understanding of the effects of food levels, abiotic factors and biotic interactions on growth, mortality and reproduction.

6.8.3 *Population dynamics of elasmobranchs*

As elasmobranchs are of much less economic importance, their population dynamics have received less attention than those of teleosts. Their life-history characteristics of longevity, slow-growth, late maturity, a long gestation period and low fecundity, suggest that their response to a change in population density by a change in recruitment will be slower and much smaller than most teleosts. Consequently, they may be less able to compensate for losses to a fishery than teleosts and only able to support light fishing pressure. In the Irish Sea, the 'common' skate, *Raia batis*, has been driven to the brink of extinction by fishing pressure (Brander, 1988). And in the Gulf of Thailand, the demersal fishery has seen the catch of rays drop from 14.8 kg per trawling hour in 1963 to 0.1 in 1982 (Pauly, 1988).

6.9 Production

Production is the total quantity of fish tissue synthesized by a population over a defined time-period, including that synthesized by individuals dying within that interval. Consequently, it is the resultant of biologically different processes: growth, mortality and reproduction. For individual freshwater species, estimates of production vary from $0.01–155.4\,\mathrm{g\,m^{-2}\,y^{-1}}$. Estimates for the ratio of production to biomass, P/B, range from 0.2–7.5 (Mann and Penczak, 1986).

Even within a population, the annual production can vary widely. Although values for production give a gross indication of the potential yields to other consumers from a population, the values themselves provide little insight into the role of the population in the functioning of the community. In aquatic communities, many interactions are dependent on the sizes of the interactors. The important problem is not so much the gross magnitude of production, as the way in which that production is partitioned between different size-classes in the population. Production by the youngest age classes, when both growth rates and mortality rates are high, often forms a high proportion of the total production. Nevertheless, studies of this aspect of production are still in their infancy.

CHAPTER SEVEN
APPLIED ECOLOGY OF FISHES

7.1 Introduction

Applied ecology is the study of the effects of man's activities on the distribution and abundance of species of interest. In many cases, the effects on fish populations are the incidental by-products of activities directed by man towards other aims. In other cases, of which fishing is the most obvious, the effects are a consequence of man's exploitation of fish populations as a resource.

The key question in applied ecology is one that is sometimes strangely overlooked: what is the effect of man's activities on the net reproductive rate of the population, R_0 (chapter 6)? If this value falls below 1, the population declines, if it is equal to 1, the population remains numerically stable over time and if it is greater than 1, the population increases. This idea has been well-expressed by Caughley (1977). The management of natural populations involves three problems: (i) the treatment of a small or declining population to raise its density (conservation); (ii) the exploitation of a population to take from it a sustained yield (harvesting); and (iii) the treatment of a population that is too dense or that has an unacceptably high rate of increase, to stabilize or reduce its density (pest control).

The mean net reproductive rate of a population is a function of its birth and death rates (chapter 6). In most fish populations, both death rates and fecundity are size-dependent (chapters 3, 6), and so growth rates are also an important factor in determining the mean net reproductive rate. Thus, a problem in applied ecology can be defined as determining the effects of man's activities on the growth, mortality and fecundity rates of the populations of interest and determining the necessary remedial actions that will have their effect by changing one or more of these rates. Complexities arise from the many factors that are involved in determining the rates of mortality, fecundity and growth in natural populations. In aquaculture, some of these complexities are by-passed by maintaining the exploited populations in simplified environments.

7.2   A classification of problems in applied ecology

With few exceptions, most of the problems of the applied ecology of fishes can be dealt with under three headings: (i) the effects of environmental degradation; (ii) the effects of fishing; and (iii) the effects of aquaculture.

7.2.1   *Environmental degradation*

This is identifiable if changes in the environment produced directly or indirectly by man's activities cause the net reproductive rate of some or all the fish populations to drop to below 1, with the consequent decline in abundance. Several forms of environmental degradation can be identified. There may be a detrimental change in the physical environment. The chemical environment experienced by the fish may become unfavourable. Organisms may be introduced into the environment that cause declines in the abundance of the fish species already present. These exotic organisms can be other fishes, or other animals including fish parasites, plants or microbial pathogens.

7.2.2   *Fishing*

Fishing is an intentional activity, although the motivation for the fishery may vary. Some fisheries are purely recreational. Subsistence and artisanal fisheries provide food for the fishermen, their families and neighbours. Commercial fisheries exploit fish populations primarily to make an economic profit.

Where the yield taken by fishing is retained by fishermen, the inevitable result is a reduction in the numerical abundance and biomass of the fished population. Sustainable fisheries depend on exploited populations showing density dependent changes in natural mortality, fecundity and growth, which compensate for the losses to the fishery. A simple model for a fishery assumes that, when not exploited, the fish population reaches the maximum biomass that can be sustained by the environment over a long period. At this maximum equilibrium biomass, B_{max}, the birth, growth and death rates just balance each other so that the biomass neither increases nor declines (upper curve in Figure 7.1).

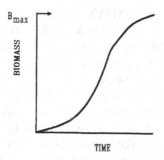

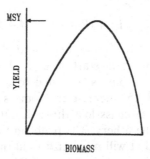

Figure 7.1 Simple model of yield to a fishery. Upper graph shows increase in biomass to the equilibrium biomass, B_{max}, of an unexploited population starting from low abundance. Lower graph shows sustainable yield that can be taken in relation to biomass present including the biomass at the maximum sustainable yield (MSY).

If the population is reduced below this equilibrium level, it responds so that the gain by births and growth now exceeds losses through deaths and in the absence of fishing, the population would increase back to its maximum, equilibrium biomass. However, this increase could also be taken by the fishery, maintaining the population at its lower, exploited biomass. In principle, the yield to the fishery can equal the increase that would occur in the absence of the fishery with the biomass of the exploited population remaining stable. If the fishery takes a greater yield, the population will decline because natural and fishing mortality exceed the natural birth and growth rates. If the fishery takes a smaller yield, the population will increase because natural and fishing mortality are less than the birth and growth rates of the exploited population. For this simple model, the sustainable yield shows a dome-shaped relationship to population biomass (lower curve in Figure 7.1). It is possible to define a maximum sustainable yield (MSY), which is the maximum yield that the fishery can take on a long-term scale while maintaining the biomass of the exploited population stable. This MSY is dependent only on the biological properties of the exploited population. It may not correspond to the yield that maximizes the profit to the fishery on a sustainable basis, the maximum sustainable economic yield because of the economics of the fishery. These will depend on non-biological quantities, including the cost of fishing and the market price of the fish (Pitcher and Hart, 1982).

The origin of the yield taken by the fishery can be seen from a simple equation originally formulated by E.S. Russell in the 1930s (Pitcher and Hart, 1982; Cushing, 1988):

$$B_2 = B_1 + (R + G) - (M + F)$$

where B_2 is the biomass of the population at the start of season 2 and B_1 is the biomass present at the start of season 1, one time period (usually a year) earlier; R is the biomass recruited to a fishable size during the time period 1 to 2; G is the increase in biomass due to growth over the same period; M is the biomass lost through natural mortality and F is the biomass taken by the fishery. The problem for the fishery manager is to chose a value of F that will maximize yield on a sustainable basis.

Fisheries managers face two problems if they set a value of F that is too high. These are growth over-fishing and recruitment over-fishing. Some success has been achieved in devising fisheries management techniques to alleviate the first (Beverton and Holt, 1957) but the second has proven more intractable and still remains a threat to many fisheries.

Growth over-fishing occurs when the intensity of fishing is so high that all the large fish are removed from the population and the fishery exploits smaller and smaller fish. Consequently, some of the growth potential of these smaller fish is not realized before they are caught. Management techniques to combat growth over-fishing include increasing the mesh-size of the nets used to catch fish so that smaller fish can escape, or reducing the intensity of fishing so at least some fish survive long enough to fulfil their growth potential (Cushing, 1988).

Analyses of the problem of growth over-fishing often assume that because the fecundity of fishes is so high, a sufficient number of fish will be recruited to the exploited population irrespective of the size of that population. This is not a realistic assumption for many fish populations. An excessively high fishing mortality, in some cases combined with unfavourable environmental conditions, can so reduce the size of the sexually mature portion of the population that the rate of reproduction is no longer sufficient to replace the numerical losses and the population declines.

To avoid recruitment over-fishing, the fishery manager must ensure that sufficient sexually mature fish are left in the population to maintain recruitment. This means that the manager must be able to predict what the recruitment from a given abundance of sexually mature fish is going to be. Unfortunately, for most fish populations it has proved difficult if not impossible to make such predictions reliably (Rothschild, 1986). This is because, in most exploited fish populations, the factors that determine the number of fish recruiting to a fishery operate in the early part of the life-history of the fish and this is a period in the life of fish that has proved most difficult to study quantitatively (chapter 6).

7.2.3 *Aquaculture*

Although modern, commercial fishing fleets use advanced technologies in the form of their boats, fishing gear and equipment for detecting fish, they remain hunters. The abundance of their prey varies with the vagaries of the environmental conditions. Aquaculture systems try to minimize the effects of environmental fluctuations by rearing fish under conditions which, to a greater or lesser extent, are controlled. Seed in the form of eggs or young fish is brought into the aquaculture system where it is grown on to a size suitable for harvesting (Figure 7.2). Aquaculture is the

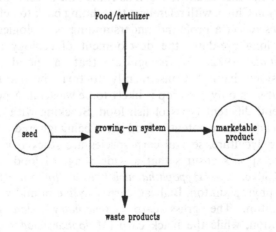

Figure 7.2 Principle of aquaculture systems. Simplified from Reay (1979).

aquatic counterpart of agriculture (Reay, 1979) and, as with agriculture, it is pursued at different levels of technology.

In the simplest, extensive aquaculture, young fish are stocked in ponds or enclosures and harvested some time later having depended on natural foods. Natural foods may be supplemented by fertilization of the ponds with organic wastes including animal and human manure, grass cuttings and other vegetable waste. In parts of southern China, mulberry trees supporting populations of silkworms are grown on the banks of the pond. The frass from the worms, pupal cocoons and tree litter drop into the ponds and fertilize the water or are eaten by the pond fish. In many developed countries, most or all of the aquaculture is intensive, with the fish being fed artificial diets often in the form of pellets such as those used on salmonid fish farms. In these intensive systems, the food usually represents a major cost to the farmer. Artificial diets are usually based on fish meal made from the catches of industrialized fisheries. There is a close linkage between the two forms of exploitation. High-quality fish protein obtained from hunting activities is used to generate equally high-quality protein as farmed fish. The justification is the difference in price that consumers are willing to pay for the farmed fish compared with the caught fish.

Many aquaculture systems are monocultures with the species kept separate throughout the growing-on period (Bardach et al., 1972; Shepherd and Bromage, 1988). However, polyculture, in which several species are grown-on together, is an important technique in China and it has now been adopted in other countries such as India and Israel. Polyculture has a long history in China, with references to it dating back to before AD 1000. Its techniques reflect a profound understanding of ecological principles, although it long pre-dates the development of ecology as a science (Bardach et al., 1972). It recognizes that a pond provides a three-dimensional habitat. Consequently, to treat the pond like a field by stocking it with only one crop is likely to be wasteful. A pond will also provide several different types of fish food. Stocking with only a single species will lead to some potential food items being ignored or under-used. In Chinese polyculture, several carp species are stocked in ponds. The choice of the species ensures that a wide range of food items is used (Figure 7.3). Silver carp (*Hypophthalmichthyes molitrix*) live in mid-water and feed on phytoplankton. Bighead carp also live in mid-water but feed on zooplankton. The grass carp (*Ctenopharyngodon idellus*) eats macrovegetation, while the black carp (*Mylopharyngodon piceus*) feeds benthically on molluscs. The omnivorous common carp may also be

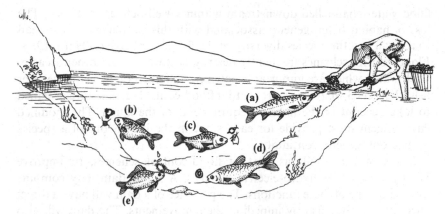

Figure 7.3 Cartoon illustrating Chinese polyculture: (a) grass carp feeding on vegetation; (b) silver carp feeding on phytoplankton; (c) bighead feeding on zooplankton; (d) black carp feeding on snails; (e) common carp feeding on benthic invertebrates.

included in this assemblage and variations on the species composition are common. The pond is fertilized by the addition of organic manures, while the grass carp are fed on grass and vegetable cuttings.

The most important cultured fishes include carps (Cyprinidae), tilapias (Cichlidae), salmon and trout (Salmonidae), eels (Anguillidae), catfish (Ictaluridae) and in coastal regions of south-east Asia, milkfish (Chanidae). The annual yield from aquaculture is about 10% of the total world yield of fishes. The yield from commercial fisheries is unlikely to increase by much but there remains great potential for further increases in yield from aquaculture, particularly low-technology systems.

7.3 Applied fish ecology of rivers

7.3.1 *Environmental degradation*

Man makes use of rivers in ways that can degrade the environment for riverine fishes. However, a deterioration in the riverine environment can also be caused by man's activities that are not directly related to the river.

7.3.1.1 *Physical structures.* These are used to modify the flow regime of the river for the purposes of flood control, navigation and power generation.

Flood-control works frequently simplify the structure of the river, with flood water channelled downstream within a well-defined river bed. The loss of habitat heterogeneity associated with this simplification can lead to a decline in the species diversity of the fish assemblage (O'Hara, 1986). Flood-control schemes frequently destroy backwaters and other areas of slack water, which are important spawning and nursery areas for the river fish (chapters 3, 6). It is important that flood-control schemes are designed to take account of the habitat requirements of the fauna. The modified environment must provide for each stage in the life history of a species to give ontogenetic continuity.

Dams may be built as part of flood-control schemes, to improve navigation and for hydro-electric schemes. A single dam may combine more than one of these functions. The presence of a dam will have a direct effect on the river fish by impeding their movements. The dam will also modify the riverine habitats and the flow regime.

A dam impedes upstream spawning-migrations (chapter 4) and can isolate a population from its normal spawning grounds. Several techniques have been developed in attempts to mitigate this. Some dams have a fish ladder constructed alongside them. This provides a corridor through which the fish can surmount the dam. The problem is to ensure that the fish can find the entrance to the ladder. A modification of the ladder is the fish lift, in which the energy to get the fish over the dam is provided by man not the fish. At some dams, the fish are transported in water tankers by road from below the dam to above it. All these techniques have been used both in North America and in Europe to allow anadromous salmonids to reach their spawning grounds. In addition, or as an alternative, fish hatcheries are built downstream of the dam so the decrease in natural recruitment caused by the loss of access to spawning grounds is compensated for by hatchery reared fish. The sexually mature fish are caught downstream of the dam and spawn in the hatchery. The juveniles are also released downstream of the dam. This technique is being used at Yichang in central China. Here, the Yoruba Dam across the Yangtse River blocks the upstreams spawning migration of sturgeon.

A dam can also cause heavy mortalities in juvenile fish migrating downstream. The fish are killed or badly damaged when they are carried over the spillway, through draw-down tunnels or through the turbines of a hydro-electric generating system.

In addition, the construction of a dam creates a lake-like environment behind it and may create a zone of turbulent tailwaters immediately below it (Neilsen et al., 1986). These changes will cause alterations in the composition of the fish assemblages. When the Volta River in West Africa was dammed to form the Volta Lake, bottom-feeding mormyrids declined but herbivorous cichlids and pelagic clupeids increased in abundance (Lowe-McConnell, 1987). The tailwaters at the base of dams built in association with locks for navigation in the Ohio and Mississippi Rivers in North America provide turbulent, oxygen-rich water and a substrate suitable for gravel-spawning fishes including the walleye (Neilson *et al.*, 1986).

7.3.1.2 *Water extraction for power plants and industrial operations.* Water extracted from the river is used for cooling by electricity generating plants and other industrial operations. The used water is usually returned to the river some degrees warmer than the unused water, thus causing a local warming of the river water.

This extraction often imposes a new cause of mortality on fish populations. The extracted water passes through screens that filter out material that would damage the plant, but these screens also capture fish drawn in with the cooling water (Wheeler, 1979).

The effect of the heated water released back into the river will vary because species in the same assemblage have different thermal tolerances (chapter 2). For eurythermal fishes, which include many cyprinid species, the heated water improves growth rates (Alabaster and Lloyd, 1982); however, cool-water fishes, including salmonids, may be excluded from the heated waters. A large plume of heated water could act as a barrier to upstream migration for such species. The zone of heated water can also be a zone of low oxygen concentrations (chapter 1).

7.3.1.3 *Mining, quarrying and forestry.* These activities can lead to physical changes in the river, which have an effect on the fish populations. Mining, quarrying and associated activities such as coal washing release large quantities of fine solids. As these settle out, they can smother the spawning grounds of salmonids and other gravel-spawning species. Extremely high levels of suspended solids can clog the gills of fishes and will eventually lead to death.

Preparation of the land for forest planting can also introduce fine solids into the water bodies. In the long-term, the forest may have a beneficial effect by damping variations in water temperature and flow rate. Nevertheless, there is also evidence that coniferous forests may amplify the effects of acidic precipitation on streams and rivers by scavenging acid-rich particles and droplets from the atmosphere (Edwards *et al.*, 1989).

7.3.1.4 *Pollution.* Poisonous substances may enter a river either by deliberate design, or inadvertently. The relatively small water volumes in rivers compared with large lakes or seas means that the adverse effects of such discharges are often seen more quickly and more dramatically in rivers. Two forms of pollution can be distinguished. There are those pollutants that have direct adverse effects on the fish by virtue of their toxicity. These include heavy metals such as cadmium, lead and zinc, other inorganic compounds including chlorine, cyanide and ammonia and organic compounds such as phenol, insecticides and herbicides. Another form of pollution is that created by organic wastes, including farmyard wastes and domestic sewage. The decomposition of these wastes by bacterial action uses up oxygen in the water. Fish are killed not by the wastes themselves but by de-oxygenation.

The pollution may occur episodically or there may be long-term chronic pollution. Because of the flushing action of the downstream flow in rivers, a recovery in water quality can be expected after episodic pollution. The ability of stream fishes to rapidly recolonize stretches that have dried out (chapter 3) suggests that the fish fauna could recover relatively quickly as conditions improve.

7.3.1.5 *Acidic precipitation.* A chronic form of pollution associated with large-scale industrialization has been recognized over the last two decades. Oxides of sulphur from the burning of sulphur-rich fossil fuels in power stations and oxides of nitrogen in emissions from vehicles have caused acid precipitation (Adams and Page, 1985). In the atmosphere, the oxides of sulphur and nitrogen become converted to acids such as sulphuric and nitric acid, which are deposited in rain and snow. In areas where the soil and underlying rock contains calcium, the acidic precipitation is neutralized. In areas where the rock and soil are poor in calcium, the run-off into streams and rivers is acidic. The effect on streams and rivers is an increase in their hydrogen-ion concentration (decrease in pH) and an increased mobilization of toxic metals including aluminium.

In Scandinavia, Scotland, Wales, eastern Canada and north-eastern United States, the decline or loss of fish populations from rivers and lakes has been correlated with increases in water acidity and metal content caused by acidic precipitation. Experimental studies have shown the adverse effects of low pH and high inorganic aluminium concentrations on fish survival and physiological performance. Unfortunately, few studies have defined the levels of pH and the related changes in metal concentrations at which the mean net reproductive rate of fish populations falls below the critical level. An exception is a study by Sadler (1983), who used a population dynamics model for brown trout in an attempt to define population effects of acidified waters in southern Norway. Such models synthesize the effects of the pollutant on survival, growth and reproductive success. Models that define the net reproductive rate under chronic levels of pollution can then be used to define the potential additional effects of episodic pollution. Episodic acidification of streams above the chronic levels occurs, particularly during the spring when snow in the river catchment is thawing. The meltwater can be highly acidic.

7.3.2   Riverine fisheries

Welcomme (1985) argues that riverine fisheries, compared with lacustrine or marine fisheries, are characterized by a diffuseness in space, their seasonality and their diversity. The diffuseness arises because the limited width of rivers means that any single landing site has too small an expanse of exploitable water to support a major capture, processing and marketing operation. Consequently, river fisheries are often located in a series of small settlements spread along the river and the fishery is labour intensive and artisanal.

Riverine fisheries are diverse. At one extreme, a fishery may be purely recreational with the fish being returned to the river after capture. In the UK recreational fishing has more participants than any other sport. At the other extreme, there may be a well-developed commercial fishery. In between, there are the subsistence fisheries pursued for limited times in the year or full-time artisanal fisheries in which the fishermen may themselves migrate to take advantage of the movements of the fish (Welcomme, 1985).

The methods of capture and the species composition of the catches are also diverse. In addition to gill nets, seines, trawls and baited hooks, which are also used in lake and sea fisheries, riverine fisheries use a variety of traps. These may be used to trap fish as flood waters recede or to capture fish as they move upstream and downstream in the river (Figure 7.4). In some rivers, the fishery is aimed at one or a few species. This is true of the fisheries that target salmonids in North America and Eurasia, but in most river systems the catch contains many different species of fish. During the development of a riverine fishery, there is a tendency for the proportion of large species in the catch to decline, so small-bodied fishes become more important—the 'fishing-down' syndrome (Welcomme, 1979, 1985).

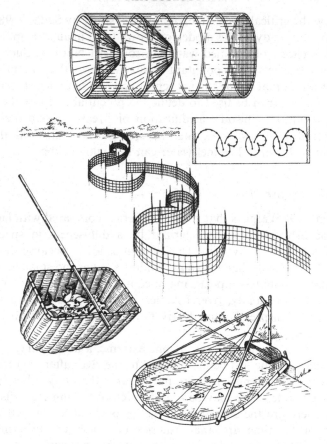

Figure 7.4 Examples of traps and nets used in riverine fisheries. Redrawn from Welcomme (1985).

7.3.3 *Aquaculture on rivers*

Many aquaculture systems are built alongside rivers to take advantage of the flowing water. Water is diverted from the river through the fish farm and then returned to the river downstream of the farm. This arrangement minimizes the expense of pumping water through the farm. A disadvantage is that waste products may be carried from the farm into the river causing some deoxygenation. Diseases may be transmitted between the cultured and wild fish populations. The fishes reared in the farm may be a different species or come from different genetic strains to the wild fish in the river. Escapees from the farm will be exotics.

Of farmed species, the stenothermal salmonids have the most demanding requirements for water quality. Production of 1 tonne of 200 g rainbow trout in water at 12°C requires a water flow of 11.1 $1s^{-1}$ (Shepherd and Bromage, 1988). A trout farm in Idaho used 4000 $1s^{-1}$ of spring water at 15°C to produce 600 000 kg per year (Bardach et al., 1972).

7.4 Applied ecology of lacustrine fishes

The long retention time of water in lakes compared with rivers means that the flushing effects are reduced, so pollutants entering a lake are likely to be retained within the lake system.

7.4.1 Environmental degradation of lakes and reservoirs

7.4.1.1 Eutrophication. Lakes are subject to many of the same forms of environmental degradation as rivers, for example inorganic and organic chemicals, sewage and heated water. In addition, lakes are probably more affected by the process of cultural eutrophication than flowing waters. Eutrophication is the enrichment of waters with plant nutrients, especially phosphorus and nitrogen. This leads to an increase in the growth of both algae and macrophytes and can result in visible plankton blooms and algal mats (Meybeck et al., 1989). The decomposition of this plant material depletes the water of oxygen and releases undesirable substances, including toxins and hydrogen sulphide. Cultural eutrophication occurs when the plant nutrients derive from man's activities. The heavy use of nitrogen and phosphorus fertilizers on agricultural land and the use both domestically and industrially of detergents containing phosphates all contribute to eutrophication.

A dramatic example of eutrophication is that of Lake Erie, one of the Great Lakes of North America. In the 1960s, the cultural eutrophication led to part of the lake (the hypolimnion of the central basin) becoming totally depleted of oxygen in the summer. An important part of the lake was unavailable to the lake fishes. These were already under pressure from the effects of over-fishing (see Section 7.4.1.3) and the other contaminants entering the lake from urban and agricultural development on the lakeshore (Christie, 1974; Meybeck et al., 1989). The effects on the water quality and the fish populations were such that the two countries bordering Lake Erie, Canada and the USA introduced a programme of waste treatment to reduce effectively the level of phosphates entering the lake.

7.4.1.2 *Lake acidification.* Lakes in watersheds with bedrock and soils that cannot buffer acidic inflows suffer from a decline in pH. A large-scale experimental study has shown how such chronic acidification can have major effects on the lake fishes (Schindler *et al.*, 1985; Mills *et al.*, 1987). Lake 223 in north-western Ontario was treated with sulphuric acid so that over the period from 1976 to 1983, its pH fell from 6.64 to 5.13. Over this period, the population densities and recruitment of several species were monitored. Over the last few years of the acidification, the species suffered recruitment failure. For several species, it was possible to estimate the threshold pH at which the failure took place: lake trout, 5.59; white sucker, 5.02; fathead minnow, 5.93; pearl dace, 5.09.

7.4.1.3 *Introduction of exotic species.* A form of environmental degradation that present evidence suggests has a more dramatic effect on lake fishes than on river fishes is the introduction of exotic species. The Great Lakes of America have experienced several such introductions and these, together with the effects of heavy fishing and environmental degradation, have led to major changes in the composition of the fish fauna (Smith, 1968; Christie, 1974; Christie *et al.*, 1987).

Until the 19th century, the Niagara Falls formed a barrier to fish movement between Lake Ontario and the other Great Lakes. With the building of the Welland Canal and subsequent improvements for navigation, a corridor linking Lakes Ontario and Erie was created. This opened up the Great Lakes to invasions by elements of the fish faunas characteristic of the St Lawrence River and even the western Atlantic. The significant invasions occurred slowly. The sea lamprey had been known in Lake Ontario in the 1830s, but was not recorded in Lake Erie until 1921, Lake Huron, 1932 and Lake Michigan, 1936. However, once established, the impact of this ectoparasitic agnathan was catastrophic. The lake trout, one of the top fish predators in the Great Lakes, had supported a commercial fishery from the 19th century. Together with other salmonid species, the trout accounted for a high proportion of the total catch until the 1940s. The trout fishery then collapsed (Figure 7.5). The cause of this collapse was probably the effect of lamprey predation on trout populations, which were already under pressure from a fishery in which the total fishing effort had been increasing. As the larger fish, the trout and the burbot, declined in abundance, the lamprey switched its attention to other salmonids—especially the lake whitefish and ciscoes (*Coregonus* spp. and *Prosopium* spp.). The larger-bodied of these species also declined in abundance, in some species to the point of extinction.

The effect of lamprey predation on the trout, burbot and larger ciscoes was to reduce the intensity of predation by these fishes on the smaller, planktivorous species. These were too small to be vulnerable to lamprey predation.

At this point, a second invader became prominent. The alewife was present in Lake Ontario but was only recorded in the other Great Lakes from the 1930s onward. This planktivorous clupeid increased greatly in abundance in Lakes Huron and Michigan following the collapse of the trout populations. Its presence was made even more obvious by its tendency to suffer periodic, massive mortalities, which littered the lake beaches with dead and rotting fish. The smallest of the lake ciscoes, the zooplanktivorous bloater, also increased in abundance with the collapse of the populations of trout and the larger ciscoes. As part of a policy to increase the density of the lake trout and other predatory fishes, a programme of lamprey control was started in the 1950s. A chemical was developed to kill the ammocoete larvae in the inlet streams in which the lampreys bred. By the late 1960s the lamprey densities had already been reduced to 10–15% of their peak numbers.

Although the changes in the American Great Lakes were not simply a result of the invasions by the lamprey and alewife (intensive fishing pressure on the commercially valuable trout also played a role), they illustrate the way in which the fish fauna of a lake can be changed by exotic species.

Two other examples emphasize this. The introduction of the Nile perch into Lake Victoria has caused the collapse of populations of cichlids, many species of which were unique to Lake Victoria (Payne, 1987; Achieng, 1990). The introduction of the predatory cichlid, *Cichla ocellaris*, into Gatun Lake in Panama caused a major reduction in the diversity and abundance of the native species (Zaret and Paine, 1973). Although exotic species have been introduced into rivers, for example a predacious percid, the zander (*Stizostedion lucioperca*) has been introduced into rivers in England from mainland Europe, the effects do not, as yet, seem to be so disastrous as in the lake examples. It remains to be discovered whether this represents a real difference in the vulnerability of lakes and rivers to invasion or is just an artefact of the examples that have been described in detail.

7.4.2 *Lake fisheries*

Lakes support recreational fisheries in many parts of the developed world but larger lakes can also support important food fisheries. The Great

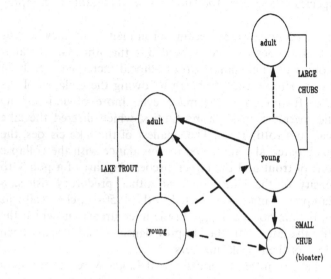

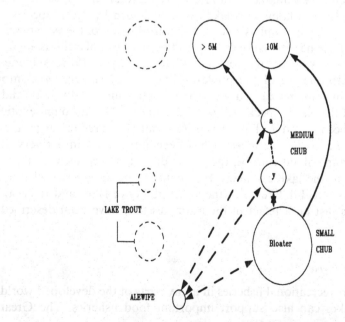

(c)   FISHERY   SEA LAMPREY   FISHERY

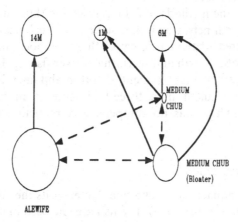

ALEWIFE

Figure 7.5 Changes in American Great Lakes fish assemblage and fisheries as a result of over-fishing and the effects of the sea lamprey. Simplified from Smith (1968). (a) unexploited stock; (b) mid 1950s; (c) mid 1960s.

Lakes of North America are an example of where recreational and commercial fisheries exist side by side. In the period 1900–1970, Lakes Ontario, Huron, Michigan and Superior yielded an annual catch that varied from 23.4 to 39.8 × 10⁶ kg. The proportion of this catch represented by salmonids declined from nearly 80% to less than 40% for reasons discussed above (Christie, 1974). However, in an effort to control the populations of alewife and bloater by increasing the density of fish predators, a policy of stocking the lakes with both the indigenous lake trout and the exotic Pacific salmons (*Oncorhynchus* spp.) has been adopted. These salmonids are also attractive to fishermen. The establishment of spawning runs of anadromous Pacific salmon in the fresh waters of the Great Lakes is a striking example of the adaptability of some teleost fishes.

In tropical lakes, subsistence and commercial fishing may continue side by side. Particularly in Africa, lake fisheries provide an important source of animal protein. Over the period from 1968 to 1988, the recorded catch from the Kenyan waters of Lake Victoria increased from 16 357 to 125 071 metric tonnes (Acheing, 1990). This increase in catch was accompanied by a major change in catch composition as the piscivorous Nile perch formed a larger and larger proportion of the total catch. Earlier, the Victoria fishery had been dominated by two indigenous cichlids,

Oreochromis esculentus and *O. variabilis*. However over-fishing, partly related to the introduction of gill nets with a reduced mesh size, caused the demise of this fishery.

In large lakes, the methods of fishing resemble those used in the open seas and include gill nets, trawls and seine nets. Technical improvements have greatly increased the efficiency of these methods and so increased the dangers of both growth and recruitment over-fishing. In the American Great Lakes fishery, the replacement of cotton and linen by nylon for gill netting material resulted in a three-fold increase in fishing intensity (Christie, 1974). There was no compensatory reduction in total fishing effort.

7.4.3 *Aquaculture in lakes*

The commonest aquaculture technique that exploits the lake environment is cage culture (Beveridge, 1987). The fish are enclosed in cages suspended in the lake (Figure 7.6). The method depends on natural water movements bringing well-oxygenated water into the cages and taking away water contaminated with the waste products from the fish. In extensive cage

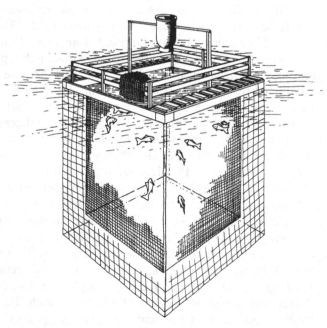

Figure 7.6 Simple system for cage culture of fish in lakes.

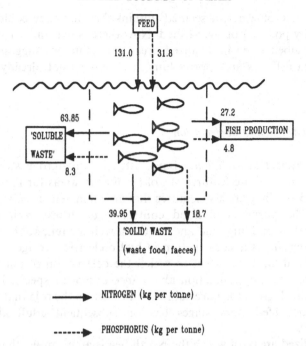

Figure 7.7 Input and outputs of nitrogen and phosphorus (kg per tonne fish produced) in cage culture of rainbow trout. Data from Beveridge (1987).

culture, the fish use natural foods generated in the lake. In intensive cage culture, the fish are fed artificial food usually in the form of pellets. Any food that is uneaten as it sinks will eventually pass out through the bottom of the cage and settle out on the substratum if not eaten by wild fish.

The waste products generated by the caged fish should not be allowed to cause a decline in the water quality in the lake that would make it unsuitable for wild fish and even for the caged fish. The waste food and waste products from the caged fish can cause a form of cultural eutrophication (Figure 7.7). In some lakes in Poland, the cage culture of rainbow trout has led to the loss of the native whitefish (Penczak *et al.*, 1982).

7.5 Applied fish ecology in the sea

For much of this century, the seas were regarded as an inexhaustible sink for waste materials and a virtually inexhaustible source of high-quality protein in the form of fishes and edible invertebrates. As the seas are such a vast volume of water, it seemed unthinkable that they could become significantly polluted or over-fished. Now there is evidence of significant pollution, albeit often in localized areas and estimates suggest that the world's sea fisheries are approaching, if they have not already reached, saturation.

7.5.1 Environmental degradation

7.5.1.1 *Physical degradation.* Bays, estuaries, mangrove swamps and other shallow, inshore features are often nursery areas for marine fishes (chapters 3, 6). They are also features that are often deliberately altered by man for the purposes of flood control, aquaculture, reclamation of agricultural or building land and, as the search for renewable sources of energy intensifies, the construction of hydro-electric barrages that make use of tidal flows. The extent to which the destruction of nursery areas has an effect on the population abundance of marine species has yet to be determined, given the uncertainties that exist on the role that mortality in the early life-history stages has on subsequent adult abundance (chapter 6).

In localized areas, of which the North Sea is an example, the extensive dredging for sand and gravel puts at risk the traditional spawning grounds of demersal spawners such as the herring (Clark, 1989). Spawning grounds may also be put at risk from the dumping of sludge derived either from the treatment of sewage on land or from dredging work in harbours and waterways. There is, as yet, little evidence linking such operations with declines in fish populations.

This lack of evidence illustrates a point that is also relevant to the chemical degradation of the marine environment and can sometimes be overlooked. Although the seas represent a vast volume of water, often small localities are crucial to the reproductive success of a fish population because they are spawning, nursery or feeding areas.

7.5.1.2 *Chemical degradation.* For many years it was assumed that any toxic substances released into the sea would be so diluted that they would have no detrimental effect on the fish populations. In recent years, the pollutants including heavy metals and organic wastes produced by long-established industries have been augmented by low-level radioactive wastes produced by the nuclear industry and oil released by large tankers or from offshore oil fields.

There is a limited capacity of shallow seas adjacent to highly industrialized countries to continually absorb and render harmless the quantities of pollutants pouring in from rivers and coastlines and entering from the atmosphere. The Baltic Sea is largely enclosed, with only a small outlet to the North Sea. Its southern coastline is well-populated and industrialized, while on the northern coastline there are major paper and wood pulp industries. The input of organic material and phosphates has led to the partial eutrophication of the Baltic. Although there has been a reduction in the relative importance of predatory fishes, evidence for deleterious effects on the offshore fisheries is lacking (Clark, 1989).

Biological processes may reverse the diluting effect of the sea. Many organisms accumulate some pollutants so the concentration in the body is higher than in the environment. This process of bioaccumulation is amplified by the process of biomagnification. This results in animals at the top of food chains accumulating even higher levels by feeding on organisms that already contain relatively high levels of the pollutant.

Some synthetic organic compounds, such as the insecticide DDT and the polychlorinated biphenyls (PCBs) used in industry, are almost insoluble in water but are soluble in lipids; thus they accumulate in fat bodies and other organs that have a high lipid content. In fish, the concentration can be 100 to 10 000 times greater than in sea water. The localization in fat reserves may mean that organic pollutants become more dangerous when the fish is under-nourished and is mobilizing its fat reserves.

Mercury is a toxic heavy metal that accumulates in fishes that are high in the food chain, such as the pelagic tunas and the demersal cod. The consumption of older tunas is discouraged because they can contain concentrations of mercury in muscle as high as 4.9 ppm. The typical value for marine fishes is about 0.15 ppm (Johnston, 1976; Clark, 1989). Mercury poisoning is one of the few clear examples where human deaths have resulted directly from marine pollution. On at least two occasions, Japanese inshore fishing communities have suffered fatalities from mercury poisoning caused by eating contaminated fish. The fish became contaminated because local industries were releasing mercury wastes into inshore waters. At Minimata, one of the sites affected, plankton contained 5 ppm mercury and fish 10–55 ppm. Acceptable levels in food fishes are about 1 ppm.

Although such local events make clear the dramatic effects that marine pollution can have, the natural variability in the abundance of marine fish populations (chapter 6) makes it difficult to detect more subtle effects.

This makes it more important that experimental studies of the effects on the components of fitness—growth, mortality and fecundity—are carried out and interpreted in terms of the effect on the mean net reproductive rate.

7.5.2 Marine fisheries

The contribution from aquaculture is increasing but sea fisheries still account for most of the global fish catch (Figure 7.8). Of the total yield of about 85 million tonnes (including invertebrates), aquaculture contributes about 12% (5% fin fish) (Shepherd and Bromage, 1988).

Marine fisheries can be classified into three basic types, coastal, upwelling and oceanic, depending on oceanographic characteristics. Coastal fisheries exploit the continental shelf zones down to a depth of 150–200 m. The main fishing techniques used are trawls, purse seines, gill nets and long lines (Figure 7.9). The catches taken include pelagic clupeoids and scombrids, demersal round fishes including gadoids, and demersal flatfishes (Figure 7.10). In temperate zones, fisheries will often concentrate on one or a few species, whereas in shelf fisheries in tropical regions such as the Gulf of Thailand, the fishery takes many species. The upwelling fisheries exploit concentrations of pelagic fishes—especially clupeoids and scombrids. These are associated with the dense plankton populations found where cool, nutrient-rich water upwells at continental margins such

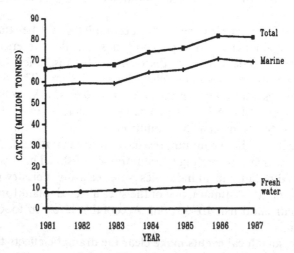

Figure 7.8 World catch of fish between 1981 and 1987 (values include yield from fisheries and aquaculture). Data from FAO Fishery Statistics, 1987.

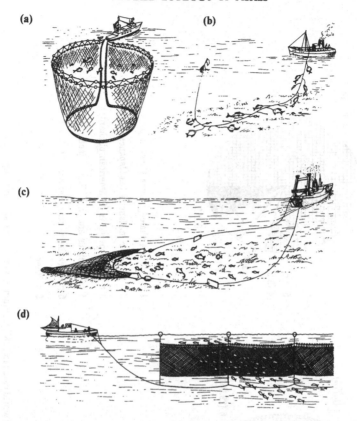

Figure 7.9 Methods used in commercial fisheries: (a) purse seining; (b) long lining; (c) trawling; (d) gill (or drift) netting.

as the coasts of Peru and California. Oceanic fisheries exploit the large pelagic carnivores, especially the tunas. They rely on long lines, drift nets or purse seines.

Although there are still many subsistence and artisanal marine fisheries, the commercial industrialized fisheries dominate. They use sophisticated boats, fishing gear and methods of locating fish. This technology has two major disadvantages for the long-term conservation of fish stocks. The first is that the high capital costs of the equipment put pressure on the fishermen to maintain the size of catches to meet these costs. Second, the methods allow fish populations to be exploited even when they are, in absolute terms, low in abundance. For example, clupeoids form schools (chapter 4). Even at low population densities, the schools will form

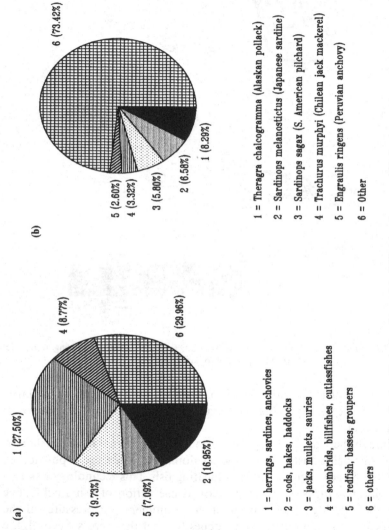

Figure 7.10   Composition of world catch in 1987: (a) main groups; (b) the six most important species. Total catch 80.837 million tonnes. Data from FAO Fishery Statistics, 1987.

localized pockets of high density. The combination of the use of sonar techniques to detect a school and the purse seine to surround the school allows a fishery to be maintained.

Brief accounts of three fisheries will illustrate some of the problems that can be confronted by those trying to develop and manage fisheries on rational principles. The problems encountered in these marine fisheries are paralleled by riverine and lacustrine fisheries.

7.5.2.1 *North Sea fishery.* The North Sea is a shallow, productive temperate sea bordered by industrialized nations. Several of these have a long tradition of commercial fisheries. The total catch is dominated by about 11 species, but the relative importance of these species has changed over time (Cole and Holden, 1973; Cushing, 1988; McIntyre, 1988). For the first 25 years after World War II, herring caught by drifting gill nets and later by mid-water trawls dominated the catch. In the mid 1960s purse seining became important. After 1973, the herring fishery collapsed and in 1977 the fishery was closed. In the 1980s the population recovered and fishing was resumed. The exploitation of another pelagic species, the mackerel, followed a similar pattern of heavy exploitation in the 1960s and collapse in the 1970s. By contrast, the exploitation of the demersal gadoids including cod, haddock and whiting increased in the 1960s and 1970s. There was also a major increase in the exploitation of the demersal sand eels and pelagic sprat. These latter species are too small to be attractive for direct human consumption but are used to produce fish meal and oil.

The collapse of the herring fishery illustrates the impact of a more effective and intense fishery on a species. Initially, the fishery takes out the older, larger fish and so becomes dependent on the younger and smaller fish. Fish are taken before or soon after they become sexually mature. The fishery becomes dependent on the reproductive success of only a small number of year-classes, each made up of small fish. It then takes only small increases in fishing pressure, perhaps combined with unfavourable climatic conditions, for reproduction and so recruitment to collapse.

It is unclear whether the increase in populations of gadoids was causally linked through predatory and competitive interactions (chapter 3) with the decline in the herring and mackerel populations, or whether the rise of demersal species and the decline of the pelagic species was coincidental.

7.5.2.2 *Peruvian upwelling fishery.* Under normal oceanographic conditions, the upwelling off the coast of Peru supports high populations of the planktivorous clupeoids, particularly the Peruvian anchoveta, and their fish and bird predators. At intervals of a few years, warm, nutrient-poor water floods southwards displacing the nutrient-rich waters. The anchoveta populations decline, normally recovering as the warm waters recede. The periodic events are called the El Nino. The anchoveta is a short-lived species so its population size tends to track closely such environmental changes.

A major Peruvian fishery for anchoveta developed in the 1950s and by the mid 1960s was the world's most important fishery with a catch of 13 million tonnes in 1970 (Cushing, 1982; Csirke, 1988). In 1972, the population and the fishery collapsed and has not subsequently regained its former abundance. This collapse was probably the result of a combination of an intense El Nino event in 1971–72 and a continuation of heavy fishing pressure during this event. Fisheries managers, at that time, adopted the conventional assumption that recruitment would remain relatively constant irrespective of the size of the population but recruitment over-fishing had occurred. A feature of this fishery was its emphasis on a single species. Ironically, on land the period of the El Nino is called the 'year of abundance' because of the rains it brings to an arid coastline (Philander, 1990).

7.5.2.3 Gulf of Thailand fishery.

A demersal trawl industry was established in this tropical sea in the 1960s (Pauly, 1988). In complete contrast to the North Sea and Peruvian fisheries, some 150 species have made a significant contribution to the fishery as it has grown from a yield of 60 thousand tonnes per year in 1960 to 800 thousand tonnes in 1980. As the fishery has grown, there have been major changes in the composition of the catch. The large, long-lived elasmobranchs (rays and sawfishes) have virtually disappeared. Their low fecundity (chapter 6) makes them vulnerable to overfishing. The larger individuals of families such as the Leiognathidae have disappeared but the proportion of the squid, an invertebrate predator, in the catches has increased.

The management of fisheries based on a small number of species is difficult because of the variations in the number of fish recruiting to the fishery in relation to the size of the adult population. The management of a fishery that includes more than 100 species, each with its own life history pattern and each subject to a myriad of factors affecting its reproductive success presents a problem whose complexity rivals that of the most complex physical systems. The development of techniques for managing multi-species fisheries represents a daunting challenge for fisheries biologists.

7.5.3 *Marine aquaculture*

Several, highly valued species are cultured in the sea using cage culture methods similar to those used in lakes. The ebb and flow of the tides and water currents flush water and some natural foods through the floating cages. Initially, the cages were sited in sheltered bays, fiords and sea lochs where there was sufficient water movement but shelter from storms. As the techniques have improved, cages that can be used in more open waters have been developed. The main centres of intensive marine cage culture are the west coasts of Canada, Ireland, Scotland and Norway and in the seas off Japan.

The North American and European centres produce salmonids, particularly the Atlantic salmon. The development of cage culture of salmon will relieve some of the fishing pressure on this valuable resource. By 1985, the annual production was 40000 tonnes (Shepherd and Bromage, 1988) out of a total yield (fishing plus culture) of 49158 tonnes (FAO, 1987). In Japan, the major farmed species are yellowtail (*Seriola quinqueradiata*), the red seabream (*Pagrus major*) and, of increasing importance, the Japanese flounder (*Paralichthys olivaceus*). In 1983, the production of marine species was 191000 tonnes.

These intensive, marine aquaculture industries all depend on the industrial fisheries for the fish meal used in the manufacture of the pelleted food. So, in a sense, the aquaculture industry is parasitic on the fishing industry rather than being a parallel development exploiting different resources. Paradoxically, the expansion of the intensive aquaculture industry, far from relieving the fishing pressure on natural populations, may increase it. This makes it even more imperative that the techniques of fisheries management and regulation continue to be improved and implemented, and that the dangers of environmental degradation to commercially exploited populations are fully evaluated. All this will depend on continued developments in our understanding of the relationship between fishes and their environment.

APPENDIX

Classification of living fishes (based on Nelson, 1984). All Orders but only selected Families relevant to the text are given.

Superclass: AGNATHA (jawless fishes)
 Order: Myxiniformes (hagfishes)
 Order: Petromyzontiformes (lampreys)
Superclass: GNATHOSTOMATA (jawed fishes)
 Class: Chondrichthyes (cartilaginous fishes)
 Subclass: Holocephali
 Order: Chimaeriformes (chimaeras)
 Subclass: Elasmobranchii (sharks and rays)
 Order: Hexanchiformes *e.g.* cow shark
   : Heterodontiformes *e.g.* Port Jackson shark
   : Lamniformes *e.g.* basking shark, mackerel shark
   : Squaliformes *e.g.* dogfish
   : Rajiformes (sharks and rays)
 Class: Osteichthyes (bony fishes)
 Subclass: Dipneusti (lungfishes)
 Order: Ceratodontiformes
   : Lepidosireniformes
 Subclass: Crossopterygii
 Order: Coelacanthiformes *e.g. Latimeria*
 Subclass: Brachiopterygii (bichirs)
 Order: Polypteriformes *e.g. Polypterus*
 Subclass: Actinopterygii (ray-finned fishes)
 Infraclass: Chondrostei
 Order: Acipenseriformes (sturgeons and paddlefish)
 Infraclass: Neopterygi 'Holosteans'
 Order: Lepisosteiformes (gars)
   : Amiiformes (bowfin)

'Teleosts'
Order: Osteoglossiformes *e.g.* Osteoglossidae, Mormyridae, Gymnar-
     chidae
    : Elopiformes *e.g.* Megalopidae (tarpons)
    : Notacanthiformes *e.g.* Halosauridae
    : Anguilliformes *e.g.* Anguillidae (eels)
    : Clupeiformes *e.g.* Clupeidae (herrings), Engraulidae (an-
     chovies)

'Superorder Ostariophysi'
Order: Gonorynchiformes *e.g.* Chanidae (milkfish)
    : Cypriniformes *e.g.* Cyprinidae (carps *etc.*), Catostomidae
     (suckers), Cobitididae (loaches)
    : Characiformes *e.g.* Curimatidae
    : Siluriformes (catfishes) *e.g.* Ictaluridae, Loricariidae,
     Ariidae
    : Gymnotiformes *e.g.* Gymnotidae

'Superorder: Protoacanthopterygii'
Order: Salmoniformes *e.g.* Salmonidae (salmon, trout, char,
     whitefish), Esocidae (pikes), Umbridae (mudminnows)

'Superorder: Stenopterygii'
Order: Stomiiformes *e.g.* Gonostomatidae, Stomiidae

'Superorder: Scopelomorpha'
Order: Aulopiformes *e.g.* Synodontidae
    : Myctophiformes *e.g.* Myctophidae (lanternfishes)

'Superorder: Paracanthopterygii'
Order: Percopsiformes *e.g.* Percopsidae (troutperches)
    : Gadiformes: Gadidae (cods), Macrouridae (rattails)
    : Ophidiiformes
    : Batrachoidiformes
    : Lophiiformes (anglerfishes) *e.g.* Ceratiidae
    : Gobiesociformes *e.g.* Gobiesocidae (clingfishes)

'Superorder: Acanthopterygii'
Order: Cyprindontiformes *e.g.* Exocoetidae (flyingfishes), Cyprino-
     dontidae (killifishes), Poeciliidae
    : Atheriniformes *e.g.* Atherinidae (silversides)
    : Lampriformes
    : Beryciformes *e.g.* Holocentridae (squirrelfishes)
    : Zeiformes *e.g.* Zeidae (dories)
    : Gasterosteiformes *e.g.* Gasterosteidae (sticklebacks)
    : Indostomiformes
    : Pegasiformes *e.g.* Pegasidae (sea moths)

: Syngnathiformes *e.g.* Syngnathidae (pipefish and seahorses)
: Dactylopteriformes
: Synbranchiformes *e.g.* Synbranchidae (swamp eels)
: Scorpaeniformes *e.g.* Scorpaenidae (rockfishes), Cottidae (sculpins)
: Perciformes *e.g.* Percidae, Cichlidae, Centrarchidae, Chaetodontidae, Pomacentridae, Scombridae, Carangidae
: Pleuronectiformes (flatfishes) *e.g.* Pleuronectidae (plaice)
: Tetraodontiformes *e.g.* Balisitidae (triggerfish), Molidae (sunfish)

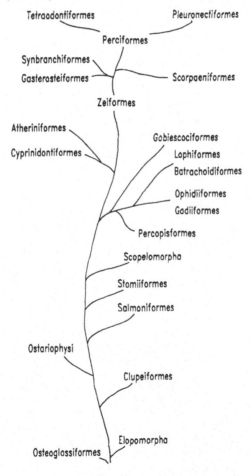

Figure A1 Simplified and provisional phylogenetic tree for teleost fishes (simplified and redrawn from Nelson, 1984).

REFERENCES

Achieng, A.P. (1990). The impact of the introduction of Nile perch, *Lates niloticus* (L.) on the fisheries of Lake Victoria. *J. Fish Biol.*, **37 (Suppl. A)**, 17–23.

Adams, D.D. and Page, W.P. (1985). *Acid Deposition, Environmental, Economic and Policy Issues.* Plenum. New York.

Alabaster, J.S. and Lloyd, R. (1982). *Water Quality Criteria for Freshwater Fish, 2nd edn.* Butterworth Scientific, Guildford.

Alm, G. (1959). Connection between maturity, size and age in fishes. *Rep. Inst. Freshwat. Res. (Drottningholm)*, **40**, 5–145.

Angermeier, P.L. and Karr, J.R. (1984). 'Fish communities along environmental gradients in a system of tropical streams' in *Evolutionary Ecology of Neotropical Freshwater Fishes*, T. M. Zaret (ed.), Dr W. Junk (publisher), The Hague, pp. 39–57.

Arnold, G.P. (1981). 'Movements of fish in relation to water currents' in *Animal Migration*, D.J. Aidley (ed.), Cambridge University Press, Cambridge, pp. 54–79.

Aronson, L.R. (1951). Orientation and jumping behaviour in the gobiid fish *Bathygobius saporator*. *Am. Mus. Novit.*, **1486**, 1–21.

Bagenal, T.B. (1978). 'Aspects of fish fecundity' in *Ecology of Fish Production*, S.D. Gerking (ed.), Blackwell, Oxford, pp. 75–101.

Bailey, K.M. and Houde, E.D. (1989). Predation on eggs and larvae of marine fishes and the recruitment problem. *Adv. Mar. Biol.*, **25**, 1–83.

Balon, E.K. (1975). Reproductive guilds of fishes: a proposal and definition. *J. Fish. Res. Bd. Can.*, **32**, 821–864.

Balon, E.K., Bruton, M.N. and Fricke, H. (1988). A fiftieth anniversary reflection on the living coelacanth, *Latimeria chalumnae*: some new interpretations of its natural history and conservation status. *Env. Biol. Fish.*, **23**, 241–280.

Bardach, J.E., Ryther, J.H. and McLarney, W.O. (1972). *Aquaculture.* Wiley-Interscience, New York.

Bayliff, W.H. (1980). Synopses of biological data on eight species of scombrids. *Inter-Am. Trop. Tuna Comm. Sp. Rep.*, **2**, 1–530.

Beamish, F.W.H. (1978). 'Swimming capacity' in *Fish Physiology, Vol. VIII*, W.S. Hoar and D.J. Randall (eds.) Academic Press, London, pp. 101–187.

Beamish, F.W.H. and Medland, T.E. (1988). Age determination in lampreys. *Trans. Am. Fish. Soc.*, **117**, 63–71.

Begon, M., Harper, J.L. and Townsend, C.R. (1989). *Ecology. Individuals, Populations and Communities (2nd edn).* Blackwell, Oxford.

Beveridge, M. (1987). *Cage Aquaculture.* Fishing News Books Ltd, Farnham.

Beverton, R.J.H. and Holt, S.J. (1957). On the dynamics of exploited fish populations. *Fishery Invest. (London)*, **19**, 533.

Bishop, J.E. (1973). *Limnology of a Small Malayan River Sungai Gombak.* Dr W. Junk (publisher), The Hague.

Blaxter, J.H.S. and Hunter, J.R. (1982). The biology of the clupeoid fishes. *Adv. Mar. Biol.*, **20**, 1–223.

Bleckmann, H. (1986). 'Role of the lateral line in fish behaviour' in *The Behaviour of Teleost Fishes*, T.J. Pitcher, (ed.), Croom Helm, London, pp. 177–202.

Bond, C.E. (1979). *Biology of Fishes*. W.B. Saunders, Philadelphia.

Bone, Q. and Marshall, N.B. (1982). *Biology of Fishes*. Blackie, Glasgow.

Bowen, S.H. (1984). 'Detritivory in neotropical fish communities' in *Evolutionary Ecology of Neotropical Freshwater Fishes*, T.M. Zaret (ed.), Dr W. Junk (publisher), The Hague, pp. 59–66.

Brafield, A.E. and Llewellyn, M.J. (1982). *Animal Energetics*. Blackie, Glasgow.

Brander, K. (1988). 'Multispecies fisheries of the Irish Sea' in *Fish Population Dynamics (2nd edn.)*, J.A. Gulland (ed.), John Wiley, Chichester, pp. 303–328.

Brandt, S.B., Magnuson, J.J. and Crowder, L.B. (1980) Thermal habitat partitioning by fishes in Lake Michigan. *Can. J. Fish. Aquat. Sci.*, **37**, 1557–1564.

Breder, C.M. Jun. and Rosen, D.E. (1966). *Modes of Reproduction in Fishes*. Natural History Press, New York.

Brett, J.R. (1971). Energetics responses of salmon to temperature. A study of some thermal relations in the physiology and freshwater ecology of the sockeye salmon (*Oncorhynchus nerka*). *Am. Zool.*, **11**, 99–113.

Brett, J.R. (1979). 'Environmental factors and growth', in *Fish Physiology, Vol. VIII*, W.S. Hoar, D.J. Randall and J.R. Brett (eds.), Academic Press, London, pp. 599–675.

Brett, J.R. (1983). 'Life energetics of sockeye salmon, *Oncorhynchus nerka*' in *Behavioural Energetics: the Cost of Survival in Vertebrates*, W.P. Aspey and S.I. Lustick (eds.), Ohio State University Press, Columbus, pp. 29–63.

Brett, J.R. and Groves, T.D.D. (1979). 'Physiological energetics' in *Fish Physiology, Vol. VIII*, W.S. Hoar, D.J. Randall and J.R. Brett (eds.), Academic Press, London, pp. 279–352.

Budker, P. (1971). *The Life of Sharks*. Weidenfeld and Nicolson, London.

Castleberry, D.T. and Cech, J.J. Jr. (1986). Physiological responses of a native and an introduced desert fish to environmental stressors. *Ecology*, **67**, 912–918.

Caughley, G. (1977). *Analysis of Vertebrate Populations*. J. Wiley, Chichester.

Childress, J.J., Taylor, S.M., Cailliet, G.M. and Price, M.H. (1980). Patterns of growth, energy utilization and reproduction in some meso- and bathypelagic fishes off Southern California. *Mar. Biol.*, **61**, 27–40.

Christie, W.J. (1974). Changes in the fish species composition of the Great Lakes. *J. Fish. Res. Bd. Can.*, **31**, 827–854.

Christie, W.J., Spangler, G.R., Loftus, K.H., Hartman, W.L., Colby, P.J., Ross, M.A. and Talhelm, D.R. (1987). A perspective on Great Lakes fish community rehabilitation. *Can. J. Fish. Aquat. Sci.*, **44 (Suppl. 2)**, 486–499.

Clark, C.W. and Levy, D.A. (1988). Diel vertical migrations by juvenile sockeye salmon and the antipredator window. *Amer. Nat.*, **131**, 271–290.

Clark, R.B. (1989). *Marine Pollution (2nd end.)*. Clarendon Press, Oxford.

Cohen, Y. and Stone, J.N. (1987). Multivariate time series analysis of the Canadian fisheries system in Lake Superior. *Can. J. Fish. Aquat. Sci.*, **44 (Suppl. 2)**, 171–181.

Cole, H.A. and Holden, M.J. (1973). 'History of the North Sea fisheries' in *North Sea Science*, E.D. Goldberg, (ed.), M.I.T. Press, Cambridge, Mass., pp. 337–360.

Colt, J. (1984). Computation of dissolved gas concentrations in water as functions of temperature, salinity and pressure. *Am. Fish. Soc. Spec. Publ.*, **14**, 154.

Conover, D.O. and Heins, S.W. (1987). Adaptive variation in environmental and genetic sex determination in a fish. *Nature*, **326**, 496–498.

Craig, J. (1987). *The Perch*. Croom Helm, London.

Csirke, J. (1988). 'Small shoaling pelagic fish stocks' in *Fish Population Dynamics (2nd edn.)*, J.A. Gulland, (ed.), Wiley-Interscience, New York, pp. 271–302.

Cushing, D.H. (1975). *Marine Ecology and Fisheries*. Cambridge University Press, Cambridge.

Cushing, D.H. (1982). *Climate and Fisheries*. Academic Press, London.

Cushing, D.H. (1988). *The Provident Sea*. Cambridge University Press, Cambridge.

Cushing, D.H. (1990). Plankton production and year-class strength in fish populations: an update of the match/mismatch hypothesis. *Adv. Mar. Biol.*, **26**, 249–293.

Dando, P.R. (1984). 'Reproduction in estuarine fish' in *Fish Reproduction: Strategies and Tactics*, G.W. Potts and R.J. Wootton (eds.), Academic Press, London, pp. 155–170.

Day, J.H., Blaber, S.J.M. and Wallace, J.H. (1981). 'Estuarine fishes' in *Estuarine Ecology*, J.H. Day (ed.), A.A. Balkema, Rotterdam, pp. 197–221.

DeVlaming, (1971). The effects of food deprivation and salinity changes on reproductive function in the estuarine gobiid fish, *Gillichthys mirabilis*. *Biol. Bull.*, **141**, 458–471.

DeVries, A.L. (1971). 'Freezing resistance in fishes' in *Fish Physiology, Vol. VI*, W.S. Hoar and D.J. Randall (eds.), Academic Press, London, pp. 157–190.

Douglas, R.H. and Djamgoz, M.B.A. (1990). *The Visual System of Fish*. Chapman and Hall, London.

Echelle, A.A. and Kornfield, I. (eds.) (1984). *Evolution of Fish Species Flocks*. University of Maine Press, Orono, Maine.

Edwards, R.W., Gee, A.S. and Stoner, J.H. (1990). *Acid Waters in Wales*. Kluwer Academic Publishers, London.

Eggers, D.M. (1977). The nature of prey selection by planktivorous fish. *Ecology*, **63**, 381–390.

Elgar, M.A. (1990). Evolutionary compromise between a few large and many small eggs: comparative evidence in teleost fish. *Oikos*, **59**, 283–287.

Elliott, J.M. (1975). Number of meals in a day, maximum weight of food consumed in a day and maximum rate of feeding for brown trout, *Salmo trutta* L. *Freshwat. Biol.*, **5**, 287–303.

Elliott, J.M. (1979). Energetics of freshwater teleosts. *Symp. Zool. Soc. Lond.*, **44**, 29–61.

Elliott, J.M. (1981). 'Some aspects of thermal stress on freshwater teleosts' in *Stress and Fish*, A.D. Pickering (ed.), Academic Press, London, pp. 209–245.

Elliott, J.M. (1985). Population regulation for different life-stages of migratory trout, *Salmo trutta* in a Lake District stream, 1966–83. *J. Anim. Ecol.*, **54**, 617–638.

Elliott, J.M. (1989). Mechanisms responsible for population regulation in young migratory trout, *Salmo trutta*. I. The critical time for survival. *J. Anim. Ecol.*, **58**, 987–1001.

Everson, I. (1984). 'Fish biology' in *Antarctic Ecology*, R.M. Laws (ed.), Academic Press, London, pp. 157–190.

F.A.O. (1981). *Atlas of the Living Resources of the Seas*. F.A.O., Rome.

Fausch, K.D. (1984). Profitable stream positions for salmonids relating specific growth rate to net energy gain. *Can. J. Zool.*, **62**, 441–451.

Fischer, E.A. (1986). 'Mating systems of simultaneously hermaphroditic serranid fishes' in *Indo-Pacific Fish Biology*, T. Uyeno, R. Arai, T. Taniuchi and K. Matsuura (eds.), Ichthyological Society of Japan, Tokyo, pp. 776–784.

Fish, J.D. and Fish, S. (1989). *A Student's Guide to the Seashore*. Unwin Hyman, London.

FitzGerald, G.J. and Whoriskey, F.G. (1985). The effects of interspecific interactions upon male reproductive success in two sympatric sticklebacks, *Gasterosteus aculeatus* and *G. wheatlandi*. *Behaviour*, **93**, 112–125.

Foester, R.E. (1968). The sockeye salmon, *Oncorhynchus nerka*. *Bull. Fish Res. Bd. Can.*, **162**, 1–442.

Fricke, R., Handermann, H., Stahlberg, S. and Peckmann, P. (1987). The compatible critical swimming speed: a new measure for the specific swimming performance of fishes. *Zool. Jb. Physiol.*, **91**, 101–111.

Fry, F.E.J. (1971). 'The effect of environmental factors on the physiology of fish' in *Fish Physiology Vol. VI*, W.S. Hoar and D.J. Randall (eds.), Academic Press, London, pp. 1–98.

Fryer, G. and Iles, T.D. (1972). *The Cichlid Fishes of the Great Lakes*, Oliver and Boyd, Edinburgh.

Furness, R.W. (1982). Competition between fisheries and seabird communities. *Adv. Mar. Biol.*, **20**, 225–307.

Gee, J.H., Tallman, R.F. and Smart, H. (1978). Reaction of some great plains fishes to progressive hypoxia. *Can. J. Zool.*, **56**, 1962–1966.

Gee, J.M. (1989). An ecological and economic review of meiofauna as food for fish. *Zool. J. Linn. Soc.*, **96**, 243–261.

Gerking, S.D. (1989). The restricted movements of fish populations. *Biol Rev.*, **34**,221–242.

Gibson, R.N. (1969). The biology and behaviour of littoral fish. *Oceanogr. Mar. Biol. Ann. Rev.*, **7**, 367–410.

Gibson, R.N. (1982). Recent studies on the biology of intertidal fishes. *Oceanogr. Mar. Biol. Ann. Rev.*, **20**, 363–414.

Gibson, R.N. (1986). 'Intertidal teleosts: life in a fluctuating environment' in *The Behaviour of Teleost Fishes*, T.J. Pitcher (ed.), Croom Helm, London, pp. 388–408.

Giller, P.S. (1984). *Community Structure and the Niche*. Chapman and Hall, London.

Giller, P.S. and Gee, J.H.R. (1987). 'The analysis of community organization: the influence of equilibrium, scale and terminology' in *Organization of Communities Past and Present*, J.H.R. Gee and P.S. Giller (eds.), Blackwell, Oxford, pp. 519–542.

Glebe, B.D. and Leggett, W.C. (1981). Latitudinal differences in energy allocation and use during the freshwater migrations of the American shad (*Alosa sapidissima*) and their life-history consequences. *Can. J. Fish. Aquat. Sci.*, **38**, 806–820.

Goldman, B. and Talbot, F.H. (1976). 'Aspects of the ecology of coral reef fishes' in *Biology and Geology of Coral Reefs III*, O.A. Jones and R. Endean (eds.), Academic Press, London, pp. 125–154.

Gordon, J.D.M. (1979). Lifestyle and phenology in deep-sea Anacanthine teleosts. *Symp. Zool. Soc. Lond.*, **44**, 327–359.

Gorman, O.T. (1987). 'Habitat segregation in an assemblage of minnows in an Ozark stream' in *Community and Evolutionary Ecology of North American Stream Fishes*, W.J. Matthews and D.C. Heins (eds.), University of Oklahoma Press, Normal, pp. 33–41.

Goulding, M. (1980). *The Fishes and the Forest: Explorations in Amazonian Natural History*. University of California Press, Berkeley.

Goulding, M., Carvalho, M.L. and Ferreira, E.G. (1988). *Rio Negro, Rich Life in Poor Water*. SPB Academic Publishing, The Hague.

Graham, J.B. (1983). 'Heat transfer' in *Fish Biomechanics*, P.W. Webb and D. Weihs (eds.), Praeger, New York, pp. 248–279.

Green, J.M. (1971). High-tide movements and homing behaviour of the tidepool sculpin *Oligocottus maculosus*. *J. Fish. Res. Bd. Can.*, **28**, 383–389.

Gross, M.R. (1984). 'Sunfish, salmon and the evolution of alternative reproductive strategies and tactics in fishes' in *Fish Reproduction: Strategies and Tactics*, G.W. Potts and R.J. Wootton (eds.), Academic Press, London, pp. 55–75.

Gross, M.R. (1987). Evolution of diadromy in fishes. *Amer. Fish. Soc. Symp.*, **1**, 14–25.

Grossman, G.D. (1982). Dynamics and oganization of a rocky inter-tidal assemblage: the persistence and resilience of taxocene structure. *Amer. Nat.*, **119**, 611–637.

Grossman, G.D. (1986). Food partitioning in a rocky intertidal fish assemblage. *J. Zool. Lond. (B)*, **1**, 317–355.

Grossman, G.D., Moyle, P.B. and Whitaker, J.O. Jr. (1982). Stochasticity in structural and functional characteristics of an Indiana stream fish assemblage: a test of community theory. *Amer. Nat.*, **120**, 423–454.

Guthrie, D.M. (1986). 'Role of vision in fish behaviour' in *The Behaviour of Teleost Fishes*, T.J. Pitcher (ed.), Croom Helm, London, pp. 75–113.

Halver, J.E. (1989) *Fish Nutrition*, (2nd edn.) Academic Press, London.

Hanlon, R.D.G. (1981). Allochthonous plant litter as a source of organic material in an oligtrophic lake (Llyn Frongoch). *Hydrobiologia*, **80**, 257–261.

Harden Jones, F.R. (1968). *Fish Migration*. Edward Arnold, London.

Harden Jones, F.R. (1981). 'Fish migration: strategy and tactics' in *Animal Migration*, D.J. Aidley (ed.), Cambridge University Press, Cambridge, pp. 139–165.

Hardisty, M.W. (1979). *The Biology of the Cyclostomes*. Chapman and Hall, London.

Hardisty, M.W. (1986). 'General introduction to lampreys' in *The Freshwater Fishes of*

Europe, Vol 1, Part 1. Petromyzontiformes, J. Holcik, (ed.), AULA-Verlag, Wiesbaden, pp. 17–83.

Hardisty, M.W. and Potter, I.C. (1971). 'The behaviour, ecology and growth of larval lampreys' in *The Biology of Lampreys*, M.W. Hardisty and I.C. Potter (eds.), Academic Press, London, pp. 85–125.

Hart, P.J.B. (1986). 'Foraging in teleost fishes' in *The Behaviour of Teleost Fishes*, T.J. Pitcher (ed.), Croom Helm, London, pp. 211–235.

Hasler, A.D. and Scholz, A.T. (1983). *Olfactory Imprinting and Homing in Salmon*. Springer-Verlag, Berlin.

Hawkins, A.D. (1986). 'Underwater sound and fish behaviour' in *The Behaviour of Teleost Fishes*, T.J. Pitcher (ed.), Croom Helm, London, pp. 114–151.

Hearn, W.E. (1987). Interspecific competition and habitat segregation among stream-dwelling trout: a review. *Fisheries*, **12**, 24–31.

Helfman, G.S. (1981). Twilight activities and temporal structure in a freshwater fish community. *Can. J. Fish. Aquat. Sci.*, **38**, 1405–1420.

Helfman, G.S. (1986). 'Fish behaviour by day, night and twilight' in *The Behaviour of Teleost Fishes*, T.J. Pitcher (ed.), Croom Helm, London, pp. 366–387.

Henderson, P.A. (1985). An approach to the prediction of temperate freshwater fish communities. *J. Fish. Biol.*, **27 (Suppl. A)**, 279–291.

Henderson, P.A. and Walker, I. (1990). Spatial organization and population density of the fish community of the litter banks within a central Amazonian blackwater stream. *J. Fish Biol.*, **37**, 401–411.

Hildrew, A.G. (1990). Fish predation and the organization of invertebrate communities in streams. *Pol. Arch. Hydrobiol.*, **37**, 95–107.

Hixon, M.A. (1980). Competitive interactions between California reef fishes of the genus *Embiotoca. Ecology*, **61**, 918–931.

Hoar, W.S. and Randall, D.J. (1978). 'Terminology to describe swimming activity in fish' in *Fish Physiology, Vol. VII*, W.S. Hoar and D.J. Randall (eds.), Academic Press, London, xiii–xiv.

Hochachka, P.W. (1980). *Living Without Oxygen*. Harvard University Press, Harvard.

Hockachka, P.W. and Somero, G.N. (1984). *Biochemical Adaptation*. Princeton University Press, Princeton.

Hoekstra, D. and Janssen, J. (1985). Non-visual feeding behaviour of the mottled sculpin, *Cottus bairdi*, in Lake Michigan. *Env. Biol. Fish.*, **12**, 111–117.

Hokanson, K.E.F. (1977). Temperature requirements of some percids and adaptations to the seasonal temperature cycle. *J. Fish. Res. Bd. Can.*, **34**, 1524–1550.

Horn, M.H. (1972). The amount of space available for marine and freshwater fishes. *Fish. Bull.*, **70**, 1295–1297.

Horn, M.H. (1989). Biology of marine herbivorous fishes. *Oceanogr. Mar. Biol. Ann. Rev.*, **27**, 167–272.

Horne, J.K. and Campana, S.E. (1989). Environmental factors influencing the distribution of juvenile groundfish in near-shore habitats off southwest Nova Scotia. *Can. J. Fish. Aquat. Sci.*, **46**, 1277–1286.

Huet, M. (1959). Profiles and biology of western European streams as related to fish management. *Trans. Am. Fish. Soc.*, **88**, 153–163.

Hughes, G.M. (1984). 'General anatomy of gills' in *Fish Physiology, Vol. XA*, W.S. Hoar and D.J. Randall (eds.), Academic Press, London, pp. 1–72.

Ibrahim, A.A. and Huntingford, F.A. (1989). The role of visual cues in prey selection in three-spined sticklebacks (*Gasterosteus aculeatus*). *Ethology*, **81**, 265–272.

Jenkins, T.M. Jr. (1969). Social structure, position choice and microdistribution of two trout species (*Salmo trutta* and *Salmo gairdneri*) resident in mountain streams. *Anim. Behav. Monogr.*, **2**, 56–123.

Johansen, K. (1970). 'Air breathing in fishes' in *Fish Physiology, Vol. IV*, W.S. Hoar and D.J. Randall (eds.), Academic Press, London, pp. 361–411.

Johannes, P.E. (1981). *Words of the Lagoon*, University of California Press, Berkeley.

Johnston, R. (1976). 'Mechanisms and problems of marine pollution in relation to commercial fisheries' in *Marine Pollution*, R. Johnston (ed.), Academic Press, London, pp. 3–156.

Jordan, D.R. and Wortley, J.S. (1985). Sampling strategy related to fish distribution, with particular reference to the Norfolk Broads. *J. Fish Biol.*, **27 (Suppl. A)**, 163–173.

Keast, A. (1978). Trophic and spatial interrelationships in the fish species of an Ontario temperate lake. *Env. Biol. Fish.*, **13**, 211–224.

Keast, A. and Webb, D. (1966). Mouth and body form relative to feeding ecology in the fish fauna of a small lake, Lake Opinicon, Ontario. *J. Fish. Res. Bd. Can.*, **23**, 1845–1874.

Kramer, D.L. (1983). The evolutionary ecology of respiratory mode in fishes: an analysis based on the cost of breathing. *Env. Biol. Fish.*, **9**, 145–158.

Krebs, C.J. (1985). *Ecology: The Experimental Analysis of Distribution and Abundance* (3rd edn) Harper and Row, New York.

Lambert, T.C. and Ware, D.M. (1984). Reproductive strategies of demersal and pelagic spawning fish. *Can. J. Fish. Aquat. Sci.*, **41**, 1565–1569.

Leggett, W.C. (1984). 'Fish migration in coastal and estuarine environments: a call for new approaches to the study of an old problem' in *Mechanisms of Migration in Fishes*, J.D. McCleave, G.P. Arnold, J.J. Dodson and W.H. Neill (eds.), Plenum, New York, pp. 159–178.

Leggett, W.C., Frank, K.T. and Carscadden, J.E. (1984). Metereological and hydrographic regulation of year-class strength in capelin (*Mallotus villosus*). *Can. J. Fish. Aquat. Sci.*, **41**, 1193–1201.

Li, H.W., Schreck, C.B., Bond, C.E. and Rexstad, E. (1987). 'Factors influencing changes in fish assemblages of Pacific Northwest streams' in *Community and Evolutionary Ecology of North American Stream Fishes*, W.J. Matthews and D.C. Heins (eds.), University of Oklahoma Press, Normal, pp. 193–202.

Lindsey, C.C. (1966). Body size of poikilotherms at different latitudes. *Evolution*, **20**, 456–465.

Linfield, R.S.J. (1985). An alternative concept to home-range theory with respect to population of cyprinids in major river systems. *J. Fish. Biol.*, **27 (Suppl. A)**, 187–196.

Longhurst, A.R. and Pauly, D. (1987). *Ecology of Tropical Oceans*. Academic Press, London.

Lotrich, V.A. (1973). Growth, production and community composition of fishes inhabiting a first-, second- and third-order stream of eastern Kentucky. *Ecol. Monogr.*, **43**, 377–397.

Lowe-McConnell, R.H. (1987). *Ecological Studies in Tropical Fish Communities*. Cambridge University Press, Cambridge.

Lundberg, J.G., Lewis, W.M. Jr, Saunders, J.F. III and Mago-Leccia, F. (1987). A major food web component in the Orinoco River channel: evidence from planktivorous fishes. *Science*, **237**, 81–83.

Lythgoe, J.N. (1979). *The Ecology of Vision*. Oxford University Press, Oxford.

McClusky, D.S. (1989). *The Estuarine Ecosystem* (2nd edn.), Blackie, Glasgow.

MacDonald, J.A., Montgomery, J.C. and Wells, R.M.G. (1987). Comparative physiology of Antarctic fishes. *Adv. Mar. Biol.*, **24**, 321–388.

McDowall, R.M. (1988). *Diadromy in Fishes*. Croom Helm, London.

McGurk, M.D. (1986). Natural mortality of marine pelagic eggs and larvae: role of spatial patchiness. *Mar. Ecol. Progr. Series*, **34**, 227–242.

McIntyre, A.D. (1988). 'Fishery resources' in *Pollution of the North Sea*, W. Salomons, B.L. Bayne, E.K. Duursma and U. Forstner (eds.), Springer-Verlag, Berlin, pp. 152–163.

McKaye, K.R. (1977). Competition for breeding sites between the cichlid fishes of Lake Jiloa, Nicaragua. *Ecology*, **58**, 291–302.

McKaye, K.R. and Gray, W.N. (1984). 'Extrinsic barriers to gene flow in rock-dwelling cichlids of Lake Malawi: macrohabitat heterogeneity and reef colonization' in *Evolution of Fish Species Flocks*, A.A. Echelle and I. Kornfield (eds.), University of Maine Press, Orano, pp. 245–273.

Mann, R.H.K. and Penczak, T. (1986). Fish production in rivers: a review. *Polskie. Arch. Hydrobiol.*, **33**, 233–247.

Marshall, N.B. (1979). *Developments in Deep-Sea Biology*. Blandford Press, Poole.

Marshall, N.B. (1984). 'Progenetic tendencies in deep-sea fishes' in *Fish Reproduction: Strategies and Tactics*, G.W. Potts and R.J. Wootton (eds.), Academic Press, London, pp. 92–101.

Marshall, T.R. and Ryan, P.A. (1987). Abundance patterns and community attributes of fishes relative to environmental gradients. *Can. J. Fish. Aquat. Sci.*, **44 (Suppl. 2)**, 198–215.

Martin, T.J. (1988). Interaction of salinity and temperature as a mechanism for spatial separation of three co-existing species of Ambassidae (Cuvier) (Teleostei) in estuaries on the south-east coast of Africa. *J. Fish Biol.*, **33 (Suppl. A)**, 9–15.

Matthews, W.J. (1987). 'Physicochemical tolerance and selectivity of stream fishes as related to their geographic ranges and local distributions' in *Community and Evolutionary Ecology of North American Stream Fishes*, W.J. Matthews and D.C. Heins (eds.), University of Oklahoma Press, Normal, pp. 111–120.

Mayr, E. (1963). *Animal Species and Their Evolution*. Harvard University Press, Harvard.

Meffe, G.K. and Sheldon, A.L. (1990). Post-defaunation recovery of fish assemblages in south-eastern blackwater streams. *Ecology*, **71**, 657–667.

Meybeck, M., Chapman, D.V. and Helmer, R. (1989). *Global Freshwater Quality*. Blackwell, Oxford.

Miller, P.J. (1979). Adaptiveness and implications of small size in teleosts. *Symp. Zool. Soc. Lond.*, **44**, 263–306.

Mills, C.A. and Mann, R.H.K. (1985). Environmentally-induced fluctuations in year-class strength and their implications for management. *J. Fish. Biol.*, **27 (Suppl. A)**, 209–226.

Mills, D. (1971). *Salmon and Trout*. Oliver and Boyd, Edinburgh.

Mills, K.H., Chalanchuk, S.M., Mohr, L.C. and Davies, I.J. (1987). Responses of fish populations in Lake 223 to 8 years of experimental acidification. *Can. J. Fish. Aquat. Sci.*, **44 (Suppl. 1)**, 114–125.

Mittelbach, G.G. (1983). Optimal foraging and growth in bluegills. *Oecologia*, **59**, 157–162.

Motta, P.J. (1988). Functional morphology of the feeding apparatus of ten species of Pacific butterflyfishes (Perciformes: Chaetodontidae): an ecomorphological approach. *Env. Biol. Fish.*, **22**, 39–67.

Moyle, P.B. and Cech, J.J. Jun. (1988). *Fishes, An Introduction to Ichthyology* (2nd edn.), Prentice Hall, Englewood Cliffs, New York.

Moyle, P.B. and Herbold, B. (1987). 'Life-history patterns and community structure in stream fishes of western North America: comparisons with eastern North America and Europe' in *Community and Evolutionary Ecology of North American Stream Fishes*, W.J. Matthews and D.C. Heins (eds.), University of Oklahoma Press, Normal, pp. 25–32.

Moyle, P.B. and Vondracek, B. (1985). Persistence and structure of the fish assemblage in a small California stream. *Ecology*, **66**, 1–13.

Mummert, J.R. and Drenner, R.W. (1986). Effect of fish size on the filtering efficiency and selective particle ingestion of a filter-feeding clupeid. *Trans. Am. Fish. Soc.*, **115**, 522–528.

Myrberg, A.A. Jun. and Thresher, R.E. (1974). Interspecific aggression and its relevance to the concept of territoriality in fishes. *Amer. Zool.*, **14**, 81–96.

Naiman, R.J. and Soltz, D.L. (eds.) (1981). *Fishes of North American Deserts*. J. Wiley, New York.

Neilson, J.D. and Perry, R.I. (1990). Diel vertical migrations of marine fishes: an obligate or facultative process? *Adv. Mar. Biol.*, **26**, 115–168.

Neilsen, L.A., Sheehan, R.J. and Orth, D.J. (1986). Impacts of navigation on riverine fish production in the United States. *Pol. Arch. Hydrobiol.*, **33**, 277–294.

Nelson, J.S. (1984). *Fishes of the World* (2nd Edn.) J. Wiley, New York.

Nikolskii, G.V. (1969). *Fish Population Dynamics*. Oliver & Boyd, Edinburgh.

Norman, J.R. (1963). *A History of Fishes* (2nd edn.), by P.H. Greenwood. Ernest Benn, London.

Northcote, T.H. (1978). 'Migratory strategies and production in freshwater fishes' in *Ecology of Freshwater Fish Production*, S.D. Gerking (ed.), Blackwell, Oxford, pp. 326–359.

O'Brien, W.J., Slade, N.A. and Vinyard, G.L. (1976). Apparent size as the determinant of prey selection by bluegill sunfish (*Lepomis macrochirus*). *Ecology*, **57**, 1304–1310.

O'Hara, K. (1986). 'Fish behaviour and the management of freshwater fisheries' in *The Behaviour of Teleost Fishes*, T.J. Pitcher (ed.), Croom Helm, London, pp. 496–521.

Odum, W.E. (1970). 'Utilization of the direct grazing and plant detritus food chains by the striped mullet, *Mugil cephalus*' in *Marine Food Chains*, J.H. Steele (ed.), Oliver and Boyd, Edinburgh, pp. 222–240.

Overholtz, W.J. and Tyler, A.V. (1986). An exploratory simulation model of competition and predation in a demersal fish assemblage on Georges Bank. *Trans. Amer. Fish. Soc.*, 115, 805–817.

Pandian, T.J. and Vivekanandan, E. (1985). 'Energetics of feeding and digestion' in *Fish Energetics New Perspectives*, P. Tytler and P. Calow (eds.), Croom Helm, London, pp. 99–124.

Pauly D. (1980). On the interrelationships between natural mortality, growth parameters and mean environmental temperature in 175 fish stocks. *J. Cons. Perm. Int. Explor. Mer.*, 39, 175–192.

Pauly, D. (1981). The relationship between gill surface area and growth performance in fish: a generalization of von Bertalanffy's theory of growth. *Meeresforschung*, 28, 251–282.

Pauly, D. (1988). 'Fisheries research and the demersal fisheries of Southeast Asia' in *Fish Population Dynamics (2nd edn.)*, J.A. Gulland, (ed.), Wiley-Interscience, New York, pp. 329–348.

Pauly, D.R. and Morgan, G.R. (1987). *Length-based Methods in Fisheries Research*. ICLARM, Manila.

Payne, A.I. (1987). A lake perched on piscine peril. *New Scient.*, 115, 50–54.

Penczak, T. (1985). Phosphorus, nitrogen and carbon cycling by fish populations in two small lowland rivers in Poland. *Hydrobiologia*, 120, 159–165.

Penczak, T., Galicka, W., Molinski, M., Kusto, E. and Zalewski, M. (1982). The enrichment of a mesotrophic lake by carbon, phosphorus and nitrogen from cage aquaculture of rainbow trout, *Salmo gairdneri*. *J. Appl. Ecol.*, 19, 371–393.

Persson, L., Andersson, G., Hamrin, S.F. and Johansson, L. (1988). 'Predator regulation and primary production along the productivity gradient of temperate lake ecosystems' in *Complex Interactions in Lake Communities*, S.R. Carpenter (ed.), Springer-Verlag, Berlin, pp. 45–65.

Philander, S.G.H. (1990). *El Nino, La Nina, and the Southern Oscillation*, Academic Press, London.

Pitcher, T.J. (1986). 'Functions of shoaling behaviour in teleosts' in *The Behaviour of Teleost Fishes*, T.J. Pitcher (ed.), Croom Helm, London, pp. 294–338.

Pitcher, T.J. and Hart, P.J.B. (1982). *Fisheries Ecology*. Croom Helm, London.

Power, M. (1987). 'Predator avoidance by grazing fishes in temperate and tropical streams: importance of stream depth and prey size' in *Predation. Direct and Indirect Effects on Aquatic Communities*, W.C. Kerfoot and A. Sih (eds.), University Press of New England, Hanover, pp. 333–351.

Power, M.E. (1990). Effects of fish in river food webs. *Science*, 250, 811–814.

Puckett, K.J. and Dill, L.M. (1984). Cost of sustained and burst swimming to juvenile coho salmon (*Oncorhynchus kisutch*). *Can. J. Fish. Aquat. Sci.*, 41, 1546–1551.

Quinn, T.P. (1984). 'An experimental approach to fish compass and map orientation' in *Mechanisms of Migration in Fishes*, J.D. McCleave, G.P. Arnold, J.J. Dodson and W.H. Neill (eds.), Plenum, New York, pp. 113–123.

Rahel, F.J. (1984). Factors structuring fish assemblages along a bog lake successional gradient. *Ecology*, 65, 1276–1289.

Reay, P.J. (1979). *Aquaculture*. Edward Arnold, London.

Reznick, D.A., Bryga, H. and Endler, J.A. (1990). Experimentally induced life-history evolution in a natural population. *Nature*, 346, 357–359.

Ricker, W.E. (1954). Stock and recruitment. *J. Fish. Res. Board Can.*, 11, 559–623.

Ricker, W.E. (1979). 'Growth rates and models' in *Fish Physiology Vol. VIII*, W.S. Hoar, D.J. Randall and J.R. Brett (eds.), Academic Press, London, pp. 677–743.

Roberts, R.J. (1989). *Fish Pathology (2nd edn.)*, Bailliere Tindall, London.

Robertson, D.R. (1990). Differences in the seasonalities of spawning and recruitment of some small neotropical reef fishes. *J. Exp. Mar. Biol. Ecol.*, **144**, 49–62.

Roff, D.A. (1981). Reproductive uncertainty and the evolution of iteroparity: why don't flatfish put all their eggs in one basket? *Can. J. Fish. Aquat. Sci.*, **38**, 968–977.

Roff, D.A. (1988). The evolution of migration and some life history parameters in marine fishes. *Env. Biol. Fish.*, **22**, 133–146.

Ross, S.T., Matthews, W.J. and Echelle, A.A. (1985). Persistence of fish assemblages: effects of environmental change. *Amer. Nat.*, **126**, 24–40.

Rothschild, B.J. (1986). *Dynamics of Marine Fish Populations*. Harvard University Press, Cambridge, Massachusetts.

Roughgarden, J. (1986). 'A comparison of food-limited and space-limited animal competition communities' in *Community Ecology*, J. Diamond and T.J. Case (eds.), Princeton University Press, Princeton, pp. 492–516.

Sadler, K. (1983). A model relating the results of low pH bioassay experiments to the fishery status of Norwegian lakes *Freshwat. Biol.*, **13**, 453–463.

Sale, P.F. (1980). The ecology of fishes on coral reefs. *Oceanogr. Mar. Biol. Ann. Rev.*, **18**, 367–421.

Sale, P.F. (1988). Perception, pattern, chance and the structure of reef communities. *Env. Biol. Fish.*, **21**, 3–15.

Sargent, R.C. and Gross, M.R. (1986). 'William's principle: an explanation of parental care in teleost fishes' in *The Behaviour of Teleost Fishes*, T.J. Pitcher (ed.), Croom Helm, London, pp. 275–293.

Schindler, D.W., Mills, K.H., Malley, D.F., Findlay, D.L., Shearer, J.A., Davies, I.J., Turner, M.A., Linsey, G.A. and Cruikshank, D.R. (1985). Long-term ecosystem stress: the effects of years of acidification on a small lake. *Science*, **28**, 1395–1401.

Schlosser, I.J. (1982). Fish community structure and function along two habitat gradients in a headwater stream. *Ecol. Monogr.*, **52**, 395–414.

Schlosser, I.J. (1987). 'A conceptual framework for fish communities in small warmwater streams' in *Community and Evolutionary Ecology of North American Stream Fishes*, W.J. Matthews and D.C. Heins (eds.), University of Oklahoma Press, Normal, pp. 17–24.

Schmidt-Nielsen, K. (1984). *Scaling: Why is Animal Size So Important?* Cambridge University Press, Cambridge.

Sedberry, G.R. and Musick, J.A. (1978). Feeding strategies of some demersal fishes of the continental shelf and rise off the mid-Atlantic coast of the USA. *Mar. Biol.*, **44**, 357–375.

Shapiro, D.Y. (1984). 'Sex reversal and sociodemographic processes in coral reef fishes' in *Fish Reproduction: Strategies and Tactics*, G.W. Potts and R.J. Wootton (eds.), Academic Press, London, pp. 103–118.

Shepherd, J. and Bromage, N. (1988). *Intensive Fish Farming*. BSP Professional Books, Oxford.

Smith, R.J.F. (1985). *The Control of Fish Migration*. Springer-Verlag, Berlin.

Smith, R.L. (1976). 'Waters of the sea: the ocean's characteristics and circulation' in *The Ecology of the Seas*, D.H. Cushing and J.J. Walsh (eds.), Blackwell, Oxford, pp. 23–58.

Smith, S.H. (1968). Species succession and fishery exploitation in the Great Lakes. *J. Fish. Res. Bd. Can.*, **25**, 667–693.

Stabell, O.B. (1984). Homing and olfaction in salmonids: a critical review with special reference to the Atlantic salmon. *Biol. Rev.*, **59**, 333–388.

Stephens, D.W. and Krebs, J.R. (1986). *Foraging Theory*. Princeton University Press, Princeton.

Stevens, E.D. and Neill, W.H. (1978). 'Body temperature relations of tunas, especially skipjack' in *Fish Physiology, Vol. VIII*, W.S. Hoar and D.J. Randall (eds.), Academic Press, London, pp. 316–359.

Stevens, J.D. (1987). *Sharks*. Merehurst Press, London.

Stott, B. (1967). The movements and population densities of roach (*Rutilus rutilus*) and gudgeon (*Gobio gobio*) in the R. Mole. *J. Anim. Ecol.*, **36**, 407–423.

Summerfeldt, R.E. and Hall, G.E. (eds.) (1987). *The Age and Growth of Fishes*. Iowa State University Press, Ames, Iowa.

Svardson, G. (1976). Interspecific population dominance in fish communities of Scandinavian lakes. *Rep. Inst. Freshwat. Res. Drottningholm*, **56**, 144–171.

Taylor, R.J. (1984). *Predation*. Chapman and Hall, London.

Thresher, R.E. (1984). *Reproduction in Reef Fishes*. T.F.H. Publications, Neptune City.

Todd, E.S. and Ebeling, A.W. (1966). Aerial respiration in the longjaw mudsucker *Gillichthys mirabilis* (Teleostei: Gobiidae). *Biol. Bull.*, **130**, 265–288.

Tonn, W.M. and Magnuson, J.J. (1982). Patterns in the species composition and richness of fish assemblages in northern Wisconsin. *Ecology*, **63**, 1149–1166.

Tonn, W.M., Paszkowski, C.A. and Moermond, T.C. (1986). Competition in *Umbra-Perca* fish assemblages: experimental and field evidence. *Oecologia*, **69**, 126–133.

Tonn, W.M., Magnuson, J.J., Rask, M. and Toivonen, J. (1990). Intercontinental comparison of small-lake fish assemblages: the balance between local and regional processes. *Amer. Nat.*, **136**, 345–375.

Trevallion, A., Steele, J.H. and Edwards, R.R.C. (1970). 'The dynamics of a benthic bivalve' in *Marine Food Chains*, J.H. Steele (ed.), Oliver and Boyd, Edinburgh, pp. 285–295.

Vanni, M.J., Luecke, C., Kitchell, J.F., Allen, Y., Temte, J. and Magnuson, J.J. (1990). Effects on lower trophic levels of massive fish mortality. *Nature*, **344**, 333–335.

Varley, M.E. (1967). *British Freshwater Fishes*. Fishing News, London.

Vogel, S. (1981). *Life in Moving Fluids*. Willard Grant Press, Boston.

Vogel, S. (1988). *Life's Devices*. Princeton University Press, Princeton.

Vrijenhoek, R.C. (1984). 'The evolution of clonal diversity in *Poeciliopsis*' in *Evolutionary Genetics of Fish*, B.J. Turner (ed.), Plenum, New York, pp. 399–429.

Vrijenhoek, R.C., Marteinsdottir, G. and Schenk, R.A. (1987). 'Genotypic and phenotypic aspects of niche diversification in fishes' in *Community and Evolutionary Ecology of North American Stream Fishes*, W.J. Matthews and D.C. Heins (eds.), University of Oklahoma Press, Normal, pp. 245–250.

Wankowski, J.W.L. and Thorpe, J.E. (1979). The role of food particle size in the growth of juvenile Atlantic salmon (*Salmo salar*). *J. Fish Biol.*, **14**, 351–370.

Warner, R.R. (1978). 'The evolution of hermaphroditism and unisexuality in aquatic and terrestrial vertebrates' in *Contrasts in Behaviour*, E.S. Reese and F.J. Lighter (eds.), Wiley Interscience, New York, pp. 78–95.

Warner, R.R. (1984). Mating behaviour and hermaphroditism. *Amer. Sci.*, **72**, 128–162.

Warner, R.R. (1988). Sex change in fishes: hypotheses, evidence and objections. *Env. Biol. Fish.*, **22**, 81–90.

Weatherley, A.H. and Gill, H.S. (1987). *The Biology of Fish Growth*. Academic Press, London.

Webb, B.W. and Walling, D.E. (1986). Spatial variation of water temperature characteristics and behaviour in a Devon river system. *Freshwat. Biol.*, **16**, 585–608.

Webb, P.W. (1975). Hydrodynamics and energetics of fish propulsion. *Bull. Fish. Res. Bd. Can.*, **190**, 1–159.

Webb, P.W. (1984). Form and function in fish swimming. *Sci. Amer.*, **251**, 58–68.

Webb, P.W. (1988). Simple physical principles and vertebrate aquatic locomotion. *Am. Zool.*, **28**, 709–725.

Webb, P.W. and Weihs, D. (1986). Functional locomotor morphology of early life history stages of fishes. *Trans. Am. Fish. Soc.*, **115**, 115–127.

Weihs, D. and Webb, P.W. (1983). 'Optimization of locomotion' in *Fish Biomechanics*, P.W. Webb and D. Weihs (eds.), Praeger, New York, pp. 339–371.

Weisberg, S.B. and Lotrich, V.A. (1986). Food limitation of a Delaware salt-marsh population of the mummichog, *Fundulus heteroclitus* (L.). *Oecologia*, **68**, 168–173.

Welcomme, R.L. (1979). *The Fisheries Ecology of Floodplain Rivers*. Longman, London.

Welcomme, R.L. (1985). River Fisheries. *FAO Tech. Paper*, **262**, 1–330.

Welcomme, R.L. (1986). 'Fish of the Niger system' in *The Ecology of River Systems*, B.R. Davies and K.F. Walker (eds.), Dr Junk (Publisher), Dordrecht, pp. 25–48.

Werner, E.E. (1984). 'The mechanisms of species interactions of community organization in fish' in *Ecological Communities: Conceptual Issues and the Evidence*, D.R. Strong, D. Simberloff, G. Abele and A.B. Thistle (eds.), Princeton University Press, Princeton, pp. 360–382.

Werner, E.E. (1986). 'Species interactions in freshwater fish communities' in *Community Ecology*, J. Diamond and T.J. Case (eds.), Princeton University Press, Princeton, pp. 344–357.

Werner, E.E. and Hall, D.J. (1974). Optimal foraging and the size selection of prey by the bluegill sunfish (*Lepomis macrochirus*). *Ecology*, **55**, 1042–1052.

Werner, E.E. and Mittelbach, G.G. (1981). Optimal foraging: field tests of diet choice and habitat switching. *Amer. Zool.*, **21**, 813–829.

Wheeler, A. (1979). *The Tidal Thames*. Routledge & Kegan Paul.

White, D.S., D'Avanzo, C., Valiela, I, Lasta, C. and Pascual, M. (1986). The relationship of diet to growth and ammonium excretion in salt marsh fish. *Env. Biol. Fish.*, **16**, 105–111.

Whoriskey, F.G. and FitzGerald, G.J. (1985). The effects of bird predation on an estuarine stickleback (Pisces: Gasterosteidae) community. *Can. J. Zool.*, **63**, 301–307.

Wikramanayake, E.D. (1990). Ecomorphology and biogeography of a tropical stream fish assemblage: evolution of assemblage structure. *Ecology*, **71**, 1756–1764.

Wildhaber, M.L. and Crowder, L.B. (1990). Testing a bioenergetics-based habitat choice model: bluegill (*Lepomis macrochirus*) responses to food availability and temperature. *Can. J. Fish. Aquat. Sci.*, **47**, 1664–1671.

Winemiller, K.O. (1990). Spatial and temporal variation in tropical fish trophic networks. *Ecol. Monogr.*, **60**, 331–367.

Winters, G.H. (1976). Recruitment mechanisms of southern Gulf of St Lawrence Atlantic herring (*Clupea harengus*). *J. Fish Res. Bd. Can.*, **33**, 1751–1763.

Witte, F. (1984). 'Ecological differentiation in Lake Victoria haplochromines: comparison of cichlid species flocks in African lakes' in *Evolution of Fish Species Flocks*, A.A. Echelle and I. Kornfield (eds.), University of Maines Press, Orano, pp. 155–167.

Woodley, J.D. *et al.* (1981). Hurricane Allen's impact on Jamaican coral reefs. *Science*, **214**, 749–755.

Wootton, R.J. (1979). Energy costs of egg production and environmental determinants of fecundity in teleost fishes. *Symp. Zool. Soc., Lond.*, **44**, 133–159.

Wootton, R.J. (1984a). *A Functional Biology of the Sticklebacks*, Croom Helm, London.

Wootton, R.J. (1984b). 'Introduction: tactics and strategies in fish reproduction' in *Fish Reproduction: Strategies and Tactics*, G.W. Potts and R.J. Wootton (eds.), Academic Press, London, pp. 1–12.

Wootton, R.J. (1990). *Ecology of Teleost Fishes*. Chapman and Hall, London.

Wourms, J.P. (1972). Developmental biology of annual fishes III. Pre-embryonic and embryonic diapause of variable duration in the eggs of annual fishes. *J. Exp. Zool.*, **182**, 389–414.

Wourms, J.P. (1991). Reproduction and development of *Sebastes* in the context of the evolution of piscine viviparity. *Env. Biol. Fish.*, **30**, 111–126.

Wourms, J.P., Grove, B.D. and Lombardi, J. (1988). 'The maternal-embryonic relationship in viviparous fishes' in *Fish Physiology Vol 11B*, W.S. Hoar and D.J. Randall (eds.), Academic Press, London, pp. 1–134.

Yang, J. (1982). An estimate of the fish biomass in the North Sea. *J. Cons. int. Explor. Mer*, **40**, 161–172.

Zalewski, M., Frankiewicz, P., Pryzybylski, M., Banbura, J. and Nowak, M. (1990). Structure and dynamics of fish communities in temperate rivers in relation to the abiotic-biotic regulatory continuum concept. *Pol. Arch. Hydrobiol.*, **37**, 151–176.

Zaret, T.M. and Paine, R.T. (1973). Species introduction in a tropical lake. *Science*, **182**, 449–455.

Zaret, T.M. and Rand, A.S. (1971). Competition in tropical stream fishes: support for the competitive exclusion principle. *Ecology*, **52**, 336–342.

Index

Printed in the United States
By Bookmasters